Lecture Notes in Computer Science 5877

Commenced Publication in 1973
Founding and Former Series Editors:
Gerhard Goos, Juris Hartmanis, and Jan van Leeuwen

Russell Deaton   Akira Suyama (Eds.)

DNA Computing and Molecular Programming

15th International Conference, DNA 15
Fayetteville, AR, USA, June 8-11, 2009
Revised Selected Papers

 Springer

Volume Editors

Russell Deaton
Dept. of Computer Science and Computer Engineering, JBHT - CSCE 504
1 University of Arkansas
Fayetteville, AR 72701, USA
E-mail: rdeaton@uark.edu

Akira Suyama
Department of Life Science and Institute of Physics
Graduate School of Arts and Sciences, The University of Tokyo
3-8-1 Komaba, Meguro-ku, Tokyo, 153-8902, Japan
E-mail: suyama@dna.c.u-tokyo.ac.jp

Library of Congress Control Number: 2009939331

CR Subject Classification (1998): F.1, F.2.2, I.2.9, J.3

LNCS Sublibrary: SL 1 – Theoretical Computer Science and General Issues

ISSN   0302-9743
ISBN-10   3-642-10603-X Springer Berlin Heidelberg New York
ISBN-13   978-3-642-10603-3 Springer Berlin Heidelberg New York

springer.com

© Springer-Verlag Berlin Heidelberg 2009
Printed in Germany

Typesetting: Camera-ready by author, data conversion by Scientific Publishing Services, Chennai, India
Printed on acid-free paper   SPIN: 12798117   06/3180   5 4 3 2 1 0

Preface

The 15th International Conference on DNA Computing and Molecular Programming was held during June 8-11, 2009 on the campus of the University of Arkansas in Fayetteville, AR. The conference attracts researchers from disparate disciplines in science and technology to foster interdisciplinary research into the molecular-scale manipulation of matter. In particular, implementation of nanoscale computation and programmed assembly of materials are of interest. Papers at the conference typically are a mix of experimental and theoretical presentations. The conference is held under the auspices of the International Society for Nanoscale Science, Computation, and Engineering (ISNSCE). The DNA15 Program Committee received 38 paper submissions, of which 20 were accepted for oral presentation and 10 for poster presentation. The meeting was attended by 68 registered participants, which included 36 students. This volume contains 16 papers selected from contributed oral presentations.

This year "Molecular Programming" was added to the title of the conference, which reflects a broader scope beyond DNA-based nanotechnology and computation. This was evident in the range of invited talks at the conference. Tutorials on "Computer Science for (not only) Molecular Biologists" by Wing-Ning Li (University of Arkansas), "Structural DNA Nanotechnology" by John Reif (Duke University), and "Autonomous Molecular Computing and Robotics" by Milan Stojanovic (Columbia University) were presented on the first day. On subsequent days, in addition to the contributed talks and posters, plenary talks were given by James Aspnes (Yale University) on "Population Protocols," Reinhard Laubenbacher (Virginia Tech) on "Discrete Model of Gene Regulation in Networks," David Leigh (University of Edinburgh) on "Synthetic Motors and Machines," Kenichi Morita (Hiroshima University) on "Computation in Reversible Cellular Automata," and Itamar Willner on "Programmed DNA assemblies for Machinery, Logic Gates, and Computing Applications."

The editors would like to thank the members of the Program Committee and the reviewers for all their hard work reviewing papers and providing constructive comments to authors. They also thank all the members of the Organizing Committee, and particularly, Jin-Woo Kim, the Co-chair of the Organizing Committee. Cindy Pickney of the Computer Science and Computer Engineering and Linda Pate of the Biological Engineering Departments at the University of Arkansas deserve special mention for all of their work. Shelly Walters at the City of Fayetteville Visitor's Bureau was an invaluable resource. They also thank all the sponsors of the conference. The editors would also like to thank the conference Steering Committee, and in particular, Natasha Jonoska, the current Chair, and Lila Kari, the previous Chair for valuable advice. Finally, the editors thank the authors, attendees, and additional staff that helped make the conference successful.

September 2009

Russell Deaton
Akira Suyama

Organization

DNA15 was organized by the Department of Computer Science and Computer Engineeering, University of Arkansas, in cooperation with the International Society for Nanoscale Science, Computation, and Engineering (ISNSCE).

Steering Committee

Natasha Jonoska	University of South Florida, USA
Leonard Adleman	University of Southern California (honorary), USA
Anne Condon	University of British Columbia, Canada
Masami Hagiya	University of Tokyo, Japan
Lika Kari	University of Western Ontario, Canada
Chengde Mao	Purdue University, USA
Giancarlo Mauri	University of Milan-Bicocca, Italy
Satoshi Murata	Tokyo Institute of Technology, Japan
John Reif	Duke University, USA
Grzegorz Rozenberg	Leiden University, The Netherlands
Nadrian Seeman	New York University, USA
Andrew Tuberfield	Oxford University, UK
Erik Winfree	California Institute of Technology, USA

Program Committee

Matteo Cavaliere	CoSBi, Trento, Italy
Anne Condon	University of British Columbia, Canada
Russell Deaton (chair)	University of Arkansas, USA
Ashish Goel	Stanford University, USA
Hendrik Jan Hoogeboom	Leiden University, The Netherlands
Lika Kari	University of Western Ontario, Canada
Ehud Keinan	Technion-Israel Institute of Technology, Israel
Thom LaBean	Duke University, USA
Giancarlo Mauri	University of Milan-Bicocca, Italy
Yongli Mi	Hong Kong University of Science and Technology
Satoshi Murata	Tokyo Institute of Technology, Japan
Andrei Paun	Louisiana Tech University, USA
Paul Rothemund	California Institute of Technology, USA
Yasu Sakakibara	Keio University, Japan
Ned Seeman	New York University, USA
Friedrich Simmel	Technische Universität München, Germany

Petr Sosik   Silesian University in Opava, Czech Republic
Akira Suyama (co-chair)   University of Tokyo, Japan
Andrew Turberfield   Oxford University, UK
Ron Weiss   Princeton University, USA
Masayuki Yamamura   Tokyo Institute of Technology, Japan
Hao Yan   Arizona State University, USA
Bernie Yurke   Boise State University, USA

Organizing Committee

Russell Deaton (Chair)   University of Arkansas
Jin-Woo Kim (Co-chair)   University of Arkansas
George Holmes   University of Arkansas
Chaim Goodman-Strauss   University of Arkansas
Wing-Ning Li   University of Arkansas

Referees

Cheng, Qi   Ibrahimi, Morteza   Masson, Benoit
Davidsohn, Noah   Iaccarino, Gennaro   Saurabh, Gupta
Gupta, Saurabh   Jack, John   Seki, Shinnosuke

Sponsoring Institutions

Air Force Office of Scientific Research
Arkansas Biosciences Institute
Arkansas Science and Technology Authority
University of Arkansas College of Engineering
University of Arkansas Graduate School
University of Arkansas Department of Computer Science and Computer
    Engineering
University of Arkansas Department of Biological Engineering
Division of Agriculture, University of Arkansas

Table of Contents

Filter Position in Networks of Evolutionary Processors Does Not Matter: A Direct Proof

Paolo Bottoni[1], Anna Labella[1], Florin Manea[2,*], Victor Mitrana[2,3,*], and Jose M. Sempere[3,**]

[1] Department of Computer Science, "Sapienza" University of Rome
Via Salaria 113, 00198 Rome, Italy
{bottoni,labella}@di.uniroma1.it
[2] Faculty of Mathematics, University of Bucharest
Str. Academiei 14, 70109 Bucharest, Romania
{flmanea,mitrana}@fmi.unibuc.ro
[3] Department of Information Systems and Computation
Technical University of Valencia,
Camino de Vera s/n. 46022 Valencia, Spain
jsempere@dsic.upv.es

Abstract. In this paper we give a direct proof of the fact that the computational power of networks of evolutionary processors and that of networks of evolutionary processors with filtered connections is the same. It is known that both are equivalent to Turing machines. We propose here a direct simulation of one device by the other. Each computational step in one model is simulated in a constant number of computational steps in the other one while a translation via Turing machines squares the time complexity.

1 Introduction

The origin of accepting networks of evolutionary processors (ANEP for short) is a basic architecture for parallel and distributed symbolic processing, related to the Connection Machine [5] as well as the Logic Flow paradigm [4], which consists of several very simple processors (called evolutionary processors), each of them being placed in a node of a virtual complete graph. By an evolutionary processor we mean an abstract processor which is able to perform very simple operations, namely point mutations in a DNA sequence (insertion, deletion or substitution of a pair of nucleotides). More generally, each node may be viewed as a cell having genetic information encoded in DNA sequences which may evolve by local evolutionary events, that is point mutations. Each node is specialized just for one of these evolutionary operations. Furthermore, the data in each

* Work supported by the PN-II Programs 11052 (GlobalComp) and 11056 (CellSim). Victor Mitrana acknowledges support from Academy of Finland, project 132727.
** Work supported by the Spanish Ministerio de Educación y Ciencia under project TIN2007-60769.

R. Deaton and A. Suyama (Eds.): DNA 15, LNCS 5877, pp. 1–11, 2009.

node is organized in the form of multisets of words (each word may appear in an arbitrarily large number of copies), and all copies are processed in parallel such that all the possible events that can take place do actually take place. Further, all the nodes send simultaneously their data and the receiving nodes handle also simultaneously all the arriving messages, according to some strategies modelled as permitting and forbidding filters and filtering criteria, see [7]. The reader interested in a more detailed discussion about the model is referred to [7,6]. In [7] one shows that this model is computationally complete and presents a characterization of the complexity class **NP** based on accepting networks of evolutionary processors (ANEP for short).

It is clear that filters associated with each node of an ANEP allow a strong control of the computation. Indeed, every node has an input and output filter; two nodes can exchange data if it passes the output filter of the sender *and* the input filter of the receiver. Moreover, if some data is sent out by some node and not able to enter any node, then it is lost. The ANEP model considered in [7] is simplified in [2] by moving the filters from the nodes to the edges. Each edge is viewed as a two-way channel such that the input and output filters, respectively, of the two nodes connected by the edge coincide. Clearly, the possibility of controlling the computation in such networks seems to be diminished. For instance, there is no possibility to lose data during the communication steps. In spite of this fact, in the aforementioned work one proves that these new devices, called accepting networks of evolutionary processors with filtered connections (ANEPFC) are still computationally complete. This means that moving the filters from the nodes to the edges does not decrease the computational power of the model. Although the two variants are equivalent from the computational power point of view, no direct proof for this equivalence has been proposed so far. It is the aim of this paper to fill this gap. We mention that both simulations presented here are time efficient, namely each computational step in one model is simulated in a constant number of computational steps in the other. This is particularly useful when one wants to translate a solution from one model into the other. A translation via a Turing machine squares the time complexity of the new solution.

2   Basic Definitions

We start by summarizing the notions used throughout the paper. An *alphabet* is a finite and nonempty set of symbols. The cardinality of a finite set A is written $card(A)$. Any finite sequence of symbols from an alphabet V is called *word* over V. The set of all words over V is denoted by V^* and the empty word is denoted by ε. The length of a word x is denoted by $|x|$ while $alph(x)$ denotes the minimal alphabet W such that $x \in W^*$.

We say that a rule $a \to b$, with $a, b \in V \cup \{\varepsilon\}$ and $ab \neq \varepsilon$ is a *substitution rule* if both a and b are not ε; it is a *deletion rule* if $a \neq \varepsilon$ and $b = \varepsilon$; it is an *insertion rule* if $a = \varepsilon$ and $b \neq \varepsilon$. The set of all substitution, deletion, and insertion rules over an alphabet V are denoted by Sub_V, Del_V, and Ins_V, respectively.

Given a rule σ as above and a word $w \in V^*$, we define the following *actions* of σ on w:

- If $\sigma \equiv a \to b \in Sub_V$, then $\sigma^*(w) = \begin{cases} \{ubv : \exists u, v \in V^* \ (w = uav)\}, \\ \{w\}, \text{ otherwise} \end{cases}$

- If $\sigma \equiv a \to \varepsilon \in Del_V$, then $\sigma^*(w) = \begin{cases} \{uv : \exists u, v \in V^* \ (w = uav)\}, \\ \{w\}, \text{ otherwise} \end{cases}$

$$\sigma^r(w) = \begin{cases} \{u : w = ua\}, \\ \{w\}, \text{ otherwise} \end{cases} \qquad \sigma^l(w) = \begin{cases} \{v : w = av\}, \\ \{w\}, \text{ otherwise} \end{cases}$$

- If $\sigma \equiv \varepsilon \to a \in Ins_V$, then

$$\sigma^*(w) = \{uav : \exists u, v \in V^* \ (w = uv)\}, \ \sigma^r(w) = \{wa\}, \ \sigma^l(w) = \{aw\}.$$

$\alpha \in \{*, l, r\}$ expresses the way of applying a deletion or insertion rule to a word, namely at any position ($\alpha = *$), in the left ($\alpha = l$), or in the right ($\alpha = r$) end of the word, respectively. For every rule σ, action $\alpha \in \{*, l, r\}$, and $L \subseteq V^*$, we define the α-*action of* σ *on* L by $\sigma^\alpha(L) = \bigcup_{w \in L} \sigma^\alpha(w)$. Given a finite set of rules M, we define the α-*action of* M on the word w and the language L by:

$$M^\alpha(w) = \bigcup_{\sigma \in M} \sigma^\alpha(w) \ \text{ and } \ M^\alpha(L) = \bigcup_{w \in L} M^\alpha(w),$$

respectively. In what follows, we shall refer to the rewriting operations defined above as *evolutionary operations* since they may be viewed as linguistic formulations of local DNA mutations.

For two disjoint subsets P and F of an alphabet V and a word w over V, we define the predicates:

$$\varphi^{(s)}(w; P, F) \equiv P \subseteq alph(w) \qquad \land \ F \cap alph(w) = \emptyset$$
$$\varphi^{(w)}(w; P, F) \equiv alph(w) \cap P \neq \emptyset \ \land \ F \cap alph(w) = \emptyset.$$

The construction of these predicates is based on *random-context conditions* defined by the two sets P (*permitting contexts/symbols*) and F (*forbidding contexts/symbols*). Informally, the first condition requires that all permitting symbols are present in w and no forbidding symbol is present in w, while the second one is a weaker variant of the first, requiring that at least one permitting symbol appears in w and no forbidding symbol is present in w. For every language $L \subseteq V^*$ and $\beta \in \{(s), (w)\}$, we define:

$$\varphi^\beta(L, P, F) = \{w \in L \mid \varphi^\beta(w; P, F)\}.$$

An *evolutionary processor over* V is a tuple (M, PI, FI, PO, FO), where:

- M is a set of substitution, deletion or insertion rules over the alphabet V. Formally: $(M \subseteq Sub_V)$ or $(M \subseteq Del_V)$ or $(M \subseteq Ins_V)$. The set M represents the set of evolutionary rules of the processor. As one can see, a processor is "specialized" in one evolutionary operation, only.
- $PI, FI \subseteq V$ are the *input* permitting/forbidding contexts of the processor, while $PO, FO \subseteq V$ are the *output* permitting/forbidding contexts of the processor. Informally, the permitting input/output contexts are the set of symbols that should be present in a word, when it enters/leaves the processor, while the

forbidding contexts are the set of symbols that should not be present in a word in order to enter/leave the processor.

We denote the set of evolutionary processors over V by EP_V. Obviously, the evolutionary processor described here is a mathematical concept similar to that of an evolutionary algorithm, both being inspired by the Darwinian evolution. The rewriting operations we have considered might be interpreted as mutations and the filtering process described above might be viewed as a selection process. Recombination is missing but it was asserted that evolutionary and functional relationships between genes can be captured by taking only local mutations into consideration [10]. Furthermore, we are not concerned here with a possible biological implementation of these processors, though a matter of great importance.

An *accepting network of evolutionary processors* (ANEP for short) is a 7-tuple $\Gamma = (V, U, G, \mathcal{N}, \alpha, \beta, x_I, x_O)$, where:

◇ V and U are the input and network alphabets, respectively, $V \subseteq U$.
◇ $G = (X_G, E_G)$ is an undirected graph, with the set of nodes X_G and the set of edges E_G. Each edge is given in the form of a binary set. G is called the *underlying graph* of the network.
◇ $\mathcal{N} : X_G \longrightarrow EP_U$ is a mapping which associates with each node $x \in X_G$ the evolutionary processor $\mathcal{N}(x) = (M_x, PI_x, FI_x, PO_x, FO_x)$.
◇ $\alpha : X_G \longrightarrow \{*, l, r\}$; $\alpha(x)$ gives the action mode of the rules of node x on the words existing in that node.
◇ $\beta : X_G \longrightarrow \{(s), (w)\}$ defines the type of the *input/output filters* of a node. More precisely, for every node, $x \in X_G$, the following filters are defined:
$$\text{input filter: } \rho_x(\cdot) = \varphi^{\beta(x)}(\cdot; PI_x, FI_x),$$
$$\text{output filter: } \tau_x(\cdot) = \varphi^{\beta(x)}(\cdot; PO_x, FO_x).$$
That is, $\rho_x(w)$ (resp. τ_x) indicates whether or not the word w can pass the input (resp. output) filter of x. More generally, $\rho_x(L)$ (resp. $\tau_x(L)$) is the set of words of L that can pass the input (resp. output) filter of x.
◇ x_I and $x_O \in X_G$ are the *input node*, and the *output node*, respectively, of the ANEP.

An *accepting network of evolutionary processors with filtered connections* (ANEPFC for short) is a 8-tuple $\Gamma = (V, U, G, \mathcal{R}, \mathcal{N}, \alpha, \beta, x_I, x_O)$, where:

◇ V, U, G, α, x_I, and x_O have the same meaning as for ANEPs.
◇ $\mathcal{R} : X_G \longrightarrow 2^{Sub_U} \cup 2^{Del_U} \cup 2^{Ins_U}$ is a mapping which associates with each node the set of evolutionary rules that can be applied in that node. Note that each node is associated only with one type of evolutionary rules, namely for every $x \in X_G$ either $\mathcal{R}(x) \subset Sub_U$ or $\mathcal{R}(x) \subset Del_U$ or $\mathcal{R}(x) \subset Ins_U$ holds.
◇ $\mathcal{N} : E_G \longrightarrow 2^U \times 2^U$ is a mapping which associates with each edge $e \in E_G$ the disjoint sets $\mathcal{N}(e) = (P_e, F_e)$, $P_e, F_e \subset U$.
◇ $\beta : E_G \longrightarrow \{s, w\}$ defines the *filter* type of an edge.

For both variants we say that $card(X_G)$ is the size of Γ.

A *configuration* of an ANEP or ANEPFC Γ as above is a mapping $C : X_G \longrightarrow 2^{V^*}$ which associates a set of words with every node of the graph. A configuration may be understood as the sets of words which are present in any node at a given moment. A configuration can change either by an *evolutionary step* or by a *communication step*.

An evolutionary step is common to both models. When changing by an evolutionary step each component $C(x)$ of the configuration C is changed in accordance with the set of evolutionary rules M_x associated with the node x and the way of applying these rules $\alpha(x)$. Formally, we say that the configuration C' is obtained in *one evolutionary step* from the configuration C, written as $C \implies C'$, if and only if

$$C'(x) = M_x^{\alpha(x)}(C(x)) \text{ for all } x \in X_G.$$

When changing by a communication step, each node processor $x \in X_G$ of an ANEP sends one copy of each word it has, which is able to pass the output filter of x, to all the node processors connected to x and receives all the words sent by any node processor connected with x provided that they can pass its input filter. Formally, we say that the configuration C' is obtained in *one communication step* from configuration C, written as $C \vdash C'$, if and only if

$$C'(x) = (C(x) - \tau_x(C(x))) \cup \bigcup_{\{x,y\} \in E_G} (\tau_y(C(y)) \cap \rho_x(C(y)))$$

for all $x \in X_G$. Note that words which leave a node are eliminated from that node. If they cannot pass the input filter of any node, they are lost.

When changing by a communication step, each node processor $x \in X_G$ of an ANEPFC sends one copy of each word it has to every node processor y connected to x, provided they can pass the filter of the edge between x and y. It keeps no copy of these words but receives all the words sent by any node processor z connected with x providing that they can pass the filter of the edge between x and z.

Formally, we say that the configuration C' is obtained in *one communication step* from configuration C, written as $C \vdash C'$, iff

$$C'(x) = (C(x) \setminus (\bigcup_{\{x,y\} \in E_G} \varphi^{\beta(\{x,y\})}(C(x), \mathcal{N}(\{x,y\}))))$$
$$\cup (\bigcup_{\{x,y\} \in E_G} \varphi^{\beta(\{x,y\})}(C(y), \mathcal{N}(\{x,y\})))$$

for all $x \in X_G$. Note that a copy of a word remains in the sending node x only if it not able to pass the filter of any edge connected to x.

Let Γ be an ANEP or ANEPFC, the computation of Γ on the input word $w \in V^*$ is a sequence of configurations $C_0^{(w)}, C_1^{(w)}, C_2^{(w)}, \ldots$, where $C_0^{(w)}$ is the initial configuration of Γ defined by $C_0^{(w)}(x_I) = \{w\}$ and $C_0^{(w)}(x) = \emptyset$ for all $x \in X_G, x \neq x_I, C_{2i}^{(w)} \implies C_{2i+1}^{(w)}$ and $C_{2i+1}^{(w)} \vdash C_{2i+2}^{(w)}$, for all $i \geq 0$. By the previous definitions, each configuration $C_i^{(w)}$ is uniquely determined by the configuration $C_{i-1}^{(w)}$. A computation as above is said to be an *accepting computation* if there exists a configuration in which the set of words existing in the output node x_O is non-empty. The *language accepted* by Γ is

$L(\Gamma) = \{w \in V^* \mid \text{ the computation of } \Gamma \text{ on } w \text{ is an accepting one}\}.$

We denote by $\mathcal{L}(ANEP)$ and $\mathcal{L}(ANEPFC)$ the class of languages accepted by ANEPs and ANEPFCs, respectively.

3   A Direct Simulation of ANEPs by ANEPFCs

Theorem 1. $\mathcal{L}(ANEP) \subseteq \mathcal{L}(ANEPFC)$.

Proof. Let $\Gamma = (V, U, G, \mathcal{N}, \alpha, \beta, x_1, x_n)$ be an ANEP with the underlying graph $G = (X_G, E_G)$ and $X_G = \{x_1, x_2, \ldots, x_n\}$ for some $n \geq 1$. We construct the ANEPFC $\Gamma' = (V, U', G', \mathcal{R}, \mathcal{N}', \alpha', \beta', x_1^s, x_n^s)$, where

$U' = U \cup \{X_i, X^d \mid X \in U, i \in \{1, \ldots, n\}\} \cup \{\$_i \mid i \in \{1, \ldots, n\}\} \cup \{\#, \$\}.$

The nodes of the graph $G' = (X_G', E_G')$, the sets of rules associated with them and the way in which they are applied, as well as the edges of E_G' together with the filters associated with them are defined in the following.

First, for every pair of nodes x_i, x_j from X_G such that $\{x_i, x_j\} \in E_G$ we have the following nodes in Γ'

- $x_{i,j}^1$: $R(x_{i,j}^1) = \{\varepsilon \to \$\}, \alpha'(x_{i,j}^1) = l$
- $x_{i,j}^2$: $R(x_{i,j}^2) = \{\$ \to \varepsilon\}, \alpha'(x_{i,j}^2) = l$,

and the following edges
- $\{x_i^f, x_{i,j}^1\}$: $P = PO(x_i), F = FO(x_i) \cup \{\$\}, \beta' = \beta(x_i)$
- $\{x_{i,j}^1, x_{i,j}^2\}$: $P = PI(x_j), F = FI(x_j), \beta' = (w)$,
- $\{x_{i,j}^2, x_j^s\}$: $P = PI(x_j), F = FI(x_i) \cup \{\$, \$_j\}, \beta' = \beta(x_j)$.

For each node x_i in Γ we add a subnetwork to Γ' according to the cases considered in the sequel.

Case 1. For an insertion or a substitution node $x_i \in X_G$ with weak filters we have the following subnetwork in Γ'.

The set of forbidden symbols F on the edge $\{x_i^f, x_i^s\}$ is defined by:
$$F = \begin{cases} FO(x_i) \cup PO(x_i) \cup \{\$_i\}, & \text{if } x_i \text{ is an insertion node} \\ PO(x_i) \cup \{\$_i\}, & \text{if } x_i \text{ is a substitution node} \end{cases}$$

Case 2. If x_i is an insertion or a substitution node with strong filters we just add the following nodes to the above construction:

$$x_i^{s,Z} : R(x_i^{s,Z}) = \{\varepsilon \to \$_i\}, \alpha'(x_i^{s,Z}) = *,$$

and the edges

$\{x_i^{s,Z}, x_i^1\} : P = \{\$_i\}, F = \{X_i \mid X \in U\}, \beta' = (w)$

$\{x_i^f, x_i^{s,Z}\} : P = U', F = \begin{cases} FO(x_i) \cup \{Z, \$_i\}, & \text{if } x_i \text{ is an insertion node} \\ \{Z, \$_i\}, & \text{if } x_i \text{ is a substitution node} \end{cases}$

$\beta' = (w)$, for all $Z \in PO(x_i)$.

Case 3. If $x_i \in X_G$ is a deletion node, then the construction in the Case 1 is modified as follows:

– The way of applying the rules in node x_i^s is changed to l if $\alpha(x_i) = r$.
– Parameters of the node x_i^1 are now $R(x_i^1) = \{X \to X^d \mid X \to \varepsilon \in M_{x_i}\}$, $\alpha'(x_i^1) = *$.
– A new node is added: x_i^4 with $R(x_i^4) = \{X^d \to \varepsilon\}$, $\alpha'(x_i^4) = \alpha(x_i)$.
– Parameters of the node x_i^f are now $R(x_i^f) = \{X^d \to X \mid X \in U\}$, $\alpha'(x_i^f) = *$.
  In this case, the edges are:
– $\{x_i^s, x_i^1\}$: $P = \{\$_i\}$, $F = \{X^d \mid X \in U\}$, $\beta' = (w)$,
– $\{x_i^1, x_i^2\}$: $P = U'$, $F = \{\#\}$, $\beta' = (w)$,
– $\{x_i^2, x_i^3\}$: $P = \{\#\}$, $F = \{\$_i\}$, $\beta' = (w)$,
– $\{x_i^3, x_i^4\}$: $P = U'$, $F = \{\#\}$, $\beta' = (w)$,
– $\{x_i^4, x_i^s\}$: $P = U'$, $F = \emptyset$, $\beta' = (w)$,
– $\{x_i^f, x_i^s\}$: $P = FO(x_i)$, $F = \{\$_i\}$, $\beta' = (w)$.

Let us follow a computation of Γ' on the input word $w \in V^*$. Let us assume that w lies in the input node x_1^s of Γ'. In the same time, we assume that w is found in x_1, the input node of Γ. Inductively, we may assume that a word w is found in some x_i, a node of Γ, as well as in x_i^s from Γ'.

In the sequel we consider two cases: x_i is a substitution or a deletion node. Since the reasoning for an insertion node is pretty similar to that for a substitution node, it is left to the reader. Let x_i be a substitution node, where a rule $Y \to X$ is applied to w producing either $w_1 X w_2$, if $w = w_1 Y w_2$ or w, if w doesn't contain Y. Here is the first difference with respect to an insertion node where every rule that could be applied is actually applied. In Γ', the word w is processed as follows. First w becomes $w\$_i$ in x_i^s, then it can enter x_i^1 only. Here it may become $w_1 X_i w_2 \$_i$, if $w = w_1 Y w_2$, or it is left unchanged. Further, $w\$_i$ can go back to x_i^s, where another $\$_i$ symbol is added to its righthand end. Then it returns to x_i^1 and the situation above is repeated. When x_i is an insertion node, then this "ping-pong" process cannot happen. On the other hand, $w_1 X_i w_2 \$_i$ enters x_i^2. It is worth mentioning that any word arriving in x_i^2 contains at most one occurrence of X_i for some $X \in U$. In x_i^2, all the symbols $\$_i$ are replaced by $\#$ which is to be deleted in x_i^3. Finally, the current word enters x_i^f where the symbol X_i, if any, is rewritten into X. Thus, in node x_i^f we have obtained the word $w_1 X w_2$, if $w = w_1 Y w_2$, or w if w doesn't contain Y; all the other words that may be obtained during these five steps either lead to the same word in x_i^f or have no effect on the rest of the computation.

The second case to be considered is when x_i is a deletion node containing a rule $Y \to \varepsilon$; we will assume that this node is a left deletion node, all the other cases being treated similarly. In this node, the word w is transformed into w', if $w = Yw'$, or is left unchanged, otherwise. In Γ' the word is processed as follows. First, in x_1^s a symbol $\$_i$ is inserted in the rightmost end of the word. Then the word enters x_i^1, where it is transformed into $w_1 Y^d w_2 \$_i$ (if $w = w_1 Y w_2$, for all the possible $w_1, w_2 \in U^*$) or $w\$_i$ (if Y doesn't occur in w). After this step, $w\$_i$ goes back to x_i^s, where another $\$_i$ symbol is added. It then returns to x_i^1 and the situation above is repeated. On the other hand, all words $w_1 Y^d w_2 \$_i$ enter x_i^2. Again, we mention that every word arriving in x_i^2 contains at most one occurrence of X^d for some $X \in U$. Here all the symbols $\$_i$ are replaced by $\#$. The words can now enter x_i^3 only, where all the symbols $\#$ are deleted. Further they go to node x_i^4, where the symbol X^d is deleted, provided that it is the leftmost one. Otherwise, they are left unchanged. Then each obtained word goes to x_i^f, where it is transformed back into w, if the symbol X^d was not deleted in the previous node, or can be left unchanged. If the word still contains X^d, then it goes back to node x_i^4 and the above considerations can be applied again. If the word obtained doesn't contain any X^d, then it is either w', where $w = Yw'$, or w; all the other words that we may obtain during these six steps either lead to the same word in x_i^f or have no effect on the rest of the computation.

In conclusion, if $w \in U^*$ is a word in the nodes x_i of Γ and x_i^s of Γ', then we can obtain $w' \in U^*$ in one processing step of Γ if and only if we can obtain w' in the node x_i^f of Γ' in 5 processing steps (if x_i is an insertion or substitution node) or in 6 processing steps (if x_i is a deletion node). At this point we note that w' can leave x_i and enters x_j in Γ if and only if w' can leave x_i^f and enters x_j^s via the nodes $x_{i,j}^1$ and $x_{i,j}^2$. If w' can leave x_i but cannot enter x_j in Γ, then it is trapped in $x_{i,j}^1$ in Γ'. Finally, if w' cannot leave node x_i, then it is resent by x_i^f to x_i^s (in the case of deletion nodes, and insertion and substitution nodes with weak filters) or to the nodes $x_i^{s,Z}$, for all $Z \in PO(x_i)$ (in the case of insertion and substitution nodes with strong filters); from this point the process described above is repeated, with the only difference that in the case of insertion and substitution nodes with strong filters, the role of node x_i^s is played by the nodes $x_i^{s,Z}$.

From the above considerations, it follows that Γ' simulates in at most 6 processing steps and 5 communication steps a processing step of Γ, and in another 2 processing steps and 3 communication steps a communication step of Γ. We conclude that $L_a(\Gamma) = L_a(\Gamma')$. $\qquad \square$

4   A Direct Simulation of ANEPFCs by ANEPs

Theorem 2. $\mathcal{L}(ANEPFC) \subseteq \mathcal{L}(ANEP)$.

Proof. Let $\Gamma = (V, U, G, \mathcal{R}, \mathcal{N}, \alpha, \beta, x_1, x_n)$ be an ANEPFC with $G = (X_G, E_G)$, X_G having n nodes $x_1, x_2, \ldots, x_n$. We construct the ANEP $\Gamma' = (V, U', G', \mathcal{N}',$

α', β', x_I, x_O), where
- $U' = V \cup X \cup \{Y\}$, $X = \{X_{i,j} \mid 1 \le i \ne j \le n, i \ne n$, and $\{x_i, x_j\} \in E_G\}$
- $G' = (X'_G, E'_G)$,
- $X'_G = \{x_I, x_O\} \cup \{x_{i,j}, x'_{i,j} \mid 1 \le i \ne j \le n, i \ne n$, and $\{x_i, x_j\} \in E_G\}$
- $E'_G = \{\{x_I, x_{1,i}\} \mid 2 \le i \le n\} \cup \{\{x_{i,j}, x'_{i,j}\} \mid 1 \le i \ne j \le n, i \ne n\} \cup$
  $\{\{x'_{i,j}, x_{j,k}\} \mid 1 \le i \ne j \le n, 1 \le j \ne k \le n\} \cup \{\{x'_{i,n}, x_O\} \mid 1 \le i \le n-1\}$,

and the other parameters defined as follows:

- node x_I: $M = \{\varepsilon \to X_{1,i} \mid 2 \le i \le n\}$,
- $PI = V$, $FI = X$, $PO = X$, $FO = \emptyset$,
- $\alpha' = *$, $\beta' = (w)$.
- nodes $x_{i,j}$, $1 \le i \ne j \le n, i \ne n$: $M = \mathcal{R}(x_i)$,
- $PI = \{X_{i,j}\}$, $FI = X \setminus \{X_{i,j}\}$, $PO = P_{\{x_i, x_j\}}$, $FO = F_{\{x_i, x_j\}}$,
- $\alpha' = \alpha(x_i)$, $\beta' = \beta(\{x_i, x_j\})$.
- nodes $x'_{i,j}$, $1 \le i \ne j \le n, i \ne n$:

$$- M = \begin{cases} \{X_{i,j} \to X_{j,k} \mid 1 \le k \le n, k \ne j\}, & \text{if } j < n \\ \{X_{i,j} \to Y\}, & \text{if } j = n \end{cases}$$

- $PI = \{X_{i,j}\}$, $FI = X \setminus \{X_{i,j}\}$, $PO = (X \cup \{Y\}) \setminus \{X_{i,j}\}$, $FO = \emptyset$,
- $\alpha' = *$, $\beta' = (w)$.
- node x_O: $M = \emptyset$, $PI = \{Y\}$, $FI = \emptyset$, $PO = \emptyset$, $FO = \emptyset$,
- $\alpha' = *$, $\beta' = (s)$.

Any computation in Γ' on an input word $w \in V^+$ produces in x_I all words $w_1 X_{1,i} w_2$ with $w_1, w_2 \in V^*$ such that $w = w_1 w_2$ and $2 \le i \le n$ provided that $\{x_1, x_i\} \in E_G$. Each word containing $X_{1,i}$ enters $x_{1,i}$. In a more general setting, we assume that a word $y_1 X_{i,j} y_2$, $y_1, y_2 \in V^*$, enters $x_{i,j}$ at a given step of the computation of Γ' on w. This means that $y = y_1 y_2$ enters x_i at a given step of the computation of Γ on w. Let y be transformed into $z = z_1 z_2$ in node x_i and z can pass the filter on the edge between x_i and x_j. Let us further assume that y_p is transformed into z_p, $p = 1, 2$. This easily implies that $y_1 X_{i,j} y_2$ is transformed into $z_1 X_{i,j} z_2$ in node $x_{i,j}$ and $z_1 X_{i,j} z_2$ can pass the output filter of $x_{i,j}$. Note that the converse is also true. Now, $z_1 X_{i,j} z_2$, $j \ne n$, enters $x'_{i,j}$ where all words $z_1 X_{j,k} z_2$, with $1 \le k \ne j \le n$ and $\{x_j, x_k\} \in E_G$, are produced. Each word $z_1 X_{j,k} z_2$ enters $x_{j,k}$ and the process of simulating the computation in Γ resumes. On the other hand, $z_1 X_{i,n} z_2$ enters $x'_{i,n}$ where $X_{i,n}$ is replaced by Y. All words produced in $x'_{i,n}$, for some $1 \le i \le n - 1$, enter x_O and the computation ends. Note that by the considerations above, a word enters $x'_{i,n}$ if and only if a word from x_i was able to pass the filter on the edge between x_i and x_n in Γ.

Note that two consecutive steps (evolutionary and communication) in Γ are simulated by four steps (two evolutionary and two communication) in Γ'. $\square$

5 Final Remarks

The simulations presented above may lead to underlying graphs of the simulating networks that differ very much from the underlying graphs of the simulated networks. However, it looks like there is some form of duality between edges and nodes in the simulations. In network theory, some types of underlying graphs are common like *rings, stars, grids*, etc. Networks of evolutionary words processors, seen as language generating or accepting devices, having underlying graphs of these special forms have been considered in several papers, see, e.g., [8] for an early survey. Simulations preserving the type of the underlying graph of the simulated network represent a matter of interest that is not considered here.

On the other hand, in almost all works reported so far ANEPs and ANEPFCs have a complete underlying graph. Starting from the observation that every ANEPFC can be immediately transformed into an equivalent ANEPFC with a complete underlying graph (the edges that are to be added are associated with filters which make them useless), we may immediately state that Theorem 1 holds for complete ANEPs and ANEPFCs as well. Although it is true that every ANEP is equivalent to a complete ANEP (every Turing machine can be simulated by a complete ANEP), we do not know a simple and direct transformation as that for ANEPFCs. Therefore, direct simulations preserving the type of the underlying graph remain to be further investigated.

The language decided by an ANEP and ANEPFC is defined in [7] and [2], respectively. It is easy to note that the construction in the proof of Theorem 2 works for a direct simulation of ANEPFCs halting on every input. However, in the proof of Theorem 1, Γ' doesn't detect the non-accepting halting computations of Γ, since configurations obtained in consecutive processing steps of Γ are not obtained here in consecutive processing steps. Thus Γ' doesn't necessarily decide the language decided by Γ. It is known from the simulation of Turing machines by ANEPs and ANEPFCs [6,3] that the languages decided by ANEPs can be also decided by ANEPFCs. Can the construction from the proof of Theorem 1 be modified for a direct simulation of ANEPs halting on every input? Finally, as the networks of evolutionary picture processors introduced in [1] do not have insertion nodes, it would be of interest to find direct simulations for these devices.

References

1. Bottoni, P., Labella, A., Mitrana, V., Sempere, J.: Networks of evolutionary picture processors (submitted)
2. Drăgoi, C., Manea, F., Mitrana, V.: Accepting networks of evolutionary processors with filtered connections. Journal of Universal Computer Science 13, 1598–1614 (2007)
3. Drăgoi, C., Manea, F.: On the descriptional complexity of accepting networks of evolutionary processors with filtered connections. International Journal of Foundations of Computer Science 19, 1113–1132 (2008)
4. Errico, L., Jesshope, C.: Towards a new architecture for symbolic processing. In: Artificial Intelligence and Information-Control Systems of Robots 1994, pp. 31–40. World Scientific, Singapore (1994)
5. Hillis, W.: The Connection Machine. MIT Press, Cambridge (1985)

6. Manea, F., Martin-Vide, C., Mitrana, V.: On the size complexity of universal accepting hybrid networks of evolutionary processors. Mathematical Structures in Computer Science 17, 753–771 (2007)
7. Margenstern, M., Mitrana, V., Perez-Jimenez, M.: Accepting hybrid networks of evolutionary systems. In: Ferretti, C., Mauri, G., Zandron, C. (eds.) DNA 2004. LNCS, vol. 3384, pp. 235–246. Springer, Heidelberg (2005)
8. Martín-Vide, C., Mitrana, V.: Networks of evolutionary processors: results and perspectives. In: Molecular Computational Models: Unconventional Approaches, pp. 78–114. Idea Group Publishing, Hershey (2005)
9. Rozenberg, G., Salomaa, A. (eds.): Handbook of Formal Languages. Springer, Berlin (1997)
10. Sankoff, D., et al.: Gene order comparisons for phylogenetic inference: evolution of the mitochondrial genome. In: Proc. Natl. Acad. Sci. USA, vol. 89, pp. 6575–6579 (1992)

Strand Algebras for DNA Computing

Luca Cardelli

Microsoft Research, 7 J J Thomson Avenue, Cambridge CB3 0FB, UK
luca@microsoft.com

Abstract. We present a process algebra for DNA computing, discussing compilation of other formal systems into the algebra, and compilation of the algebra into DNA structures.

1 Introduction

DNA technology is reaching the point where one can envision automatically compiling high-level formalisms to DNA computational structures [18]. Examples so far include the 'manual compilation' of automata and Boolean networks, where some impressive demonstrations have been carried out [18],[15],[16]. Typically one considers sequential or functional computations, realized by massive numbers of molecules; we should strive, however, to take more direct advantage of massive concurrency at the molecular level. To that end it should be useful to consider *concurrent* high-level formalism, in addition to sequential ones. In this paper we describe three compilation processes for concurrent languages. First, we compile a low-level combinatorial algebra to a certain class of composable DNA structures [17]: this is intended to be a direct (but not quite trivial) mapping, which provides an algebraic notation for writing concurrent molecular programs. Second, we compile a higher-level expression-based algebra to the low-level combinatorial algebra, as a paradigm for compiling expressions of arbitrary complexity to 'assembly language' DNA combinators.

Third is our original motivation: translating heterogeneous collections of interacting automata [4] to molecular structures. How to do that was initially unclear, because one must choose some suitable 'programmable matter' (such as DNA) as a substrate, but must also come up with compositional protocols for interaction of the components that obey the high-level semantics of the language. We show a solution to this problem in Section 5, based on the combinatorial DNA algebra. The general issue there is how to realize the *external choice* primitive of interacting automata (also present in most process algebras and operating systems), for which there is currently no direct DNA implementation. In DNA we can instead implement a *join* primitive, based on [17]: this is a powerful operator, widely studied in concurrency theory [7],[13], which can indirectly provide an implementation of external choice. The DNA algebra supporting the translation is built around the join operator.

2 Strand Algebras

By a strand algebra we mean a process algebra [11] where the main components represent DNA strands, DNA gates, and their interactions. We begin with a nondeterministic algebra, and we discuss a stochastic variant in Section 4. Our strand algebras may look very similar to either chemical reactions, or Petri nets, or multiset-rewriting systems. The difference is that the equivalent of, respectively, reactions, transitions, and rewrites, do not live *outside* the system, but rather are part of the system itself and are *consumed* by their own activity, reflecting their DNA implementation. A process algebra formulation is particularly appropriate for such an internal representation of active elements.

2.1 The Combinatorial Strand Algebra, $\mathcal{P}$

Our basic strand algebra has some atomic elements (*signals* and *gates*), and only two combinators: *parallel (concurrent) composition* P | Q, and *populations* P*. An inexhaustible population P* has the property that P* = P | P*; that is, there is always one more P that can be taken from the population. The set $\mathcal{P}$ is formally the set of finite trees P generated by the syntax below; we freely use parentheses when representing these trees linearly as strings. Up to the algebraic equations described below, each P is a multiset, i.e., a solution. The *signals* x are taken from a countable set.

2.1.1 Syntax

$$P ::= x \; \vdots \; [x_1,..,x_n].[x'_1,..,x'_m] \; \vdots \; 0 \; \vdots \; P_1 | P_2 \; \vdots \; P* \qquad n \geq 1, m \geq 0$$

A *gate* is an operator from signals to signals: $[x_1,..,x_n].[x'_1,..,x'_m]$ is a gate that binds signals $x_1..x_n$, produces signals $x'_1,..,x'_m$, and is consumed in the process. We say that this gate *joins* n signals and then *forks* m signals; see some special cases below. An inert component is indicated by 0. Signals and gates can be combined into a 'soup' by parallel composition $P_1 | P_2$ (a commutative and associative operator, similar to chemical '+'), and can also be assembled into inexhaustible populations, P*.

2.1.2 Explanation of the Syntax and Abbreviations

x		is a *signal*	0	is *inert*
$x_1.x_2$	$\overset{\text{def}}{=} [x_1].[x_2]$	is a *transducer gate*	$P_1	P_2$	is *parallel composition*
$x.[x_1,..,x_m]$	$\overset{\text{def}}{=} [x].[x_1,..,x_m]$	is a *fork gate*	P*	is *unbounded population*
$[x_1,..,x_n].x$	$\overset{\text{def}}{=} [x_1,..,x_n].[x]$	is a *join gate*		

The relation $\equiv \; \subseteq \mathcal{P} \times \mathcal{P}$, called *mixing*, is the smallest relation satisfying the following properties; it is a substitutive equivalence relation axiomatizing a well-mixed solution [2]:

2.1.3 Mixing

$P \equiv P$	equivalence	$P \equiv Q \Rightarrow P \mid R \equiv Q \mid R$	in context
$P \equiv Q \Rightarrow Q \equiv P$		$P \equiv Q \Rightarrow P^* \equiv Q^*$	
$P \equiv Q, Q \equiv R \Rightarrow P \equiv R$			
		$P^* \equiv P^* \mid P$	population
$P \mid 0 \equiv P$	diffusion	$0^* \equiv 0$	
$P \mid Q \equiv Q \mid P$		$(P \mid Q)^* \equiv P^* \mid Q^*$	
$P \mid (Q \mid R) \equiv (P \mid Q) \mid R$		$P^{**} \equiv P^*$	

The relation $\rightarrow \subseteq \boldsymbol{\mathcal{P}} x \boldsymbol{\mathcal{P}}$, called *reaction*, is the smallest relations satisfying the following properties. In addition, $\rightarrow^*$, *reaction sequence*, is the symmetric and transitive closure of $\rightarrow$.

2.1.4 Reaction

$x_1 \mid .. \mid x_n \mid [x_1,..,x_n].[x'_1,..,x'_m] \;\rightarrow\; x'_1 \mid .. \mid x'_m$	gate $(n \geq 1, m \geq 0)$
$P \rightarrow Q \;\Rightarrow\; P \mid R \rightarrow Q \mid R$	dilution
$P \equiv P', \; P' \rightarrow Q', \; Q' \equiv Q \;\Rightarrow\; P \rightarrow Q$	well mixing

The first reaction (gate) forms the core of the semantics: the other rules allow reactions to happen in context. Note that the special case of the gate rule for m=0 is $x_1 \mid ..$ $\mid x_n \mid [x_1,..,x_n].[] \;\rightarrow\; 0$. And, in particular, x.[] annihilates an x signal. We can choose any association of operators in the formal gate rule: because of the associativity of parallel composition under $\equiv$ the exact choice is not important. Since $\rightarrow$ is a relation, reactions are in general nondeterministic. Some examples are:

$x_1 \mid x_1.x_2 \;\rightarrow\; x_2$
$x_1 \mid x_1.x_2 \mid x_2.x_3 \;\rightarrow^* \; x_3$
$x_1 \mid x_2 \mid [x_1,x_2].x_3 \;\rightarrow\; x_3$
$x_1 \mid x_1.x_2 \mid x_1.x_3 \;\rightarrow\; x_2 \mid x_1.x_3 \quad \text{and} \quad \rightarrow x_3 \mid x_1.x_2$
$X \mid ([X,x_1].[x_2,X])^*$ a catalytic system ready to transform
        multiple x_1 to x_2, with catalyst X

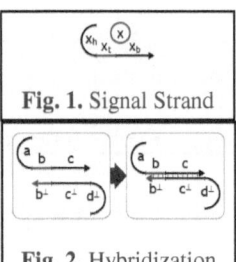

Fig. 1. Signal Strand

Fig. 2. Hybridization

There is a duality between signals and gates: signals can interact with gates but signals cannot interact with signals, nor gates with gates. As we shall see, in the DNA implementation the input part of a gate is the Watson-Crick dual of the corresponding signal strand. This duality need not be exposed in the syntax: it is implicit in the separation between signals and gates, so we use the same x_1 both for the 'positive' signal strand and for the complementary 'negative' gate input in a reaction like $x_1 \mid$ $x_1.x_2 \rightarrow x_2$.

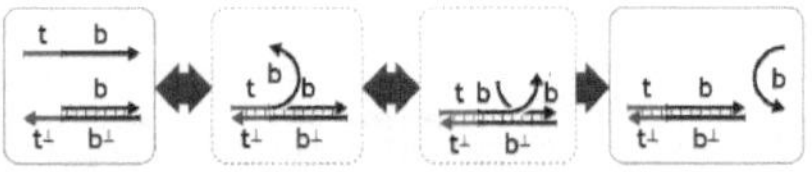

Fig. 3. Strand Displacement

3 DNA Semantics

In this section we provide a DNA implementation of the combinatorial strand algebra. Given a representation of signals and gates, it is then a simple matter to represent any strand algebra expression as a DNA system, since 0, $P_1 \mid P_2$, and P* are assemblies of signals and gates.

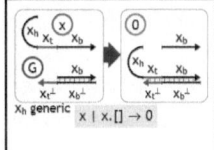

Fig. 4. Annihilator

There are many possible ways of representing signals and gates as DNA structures. First one must choose an overall architecture, which is largely dictated by a representation of signals, and then one must implement the gates, which can take many forms with various qualitative and quantitative trade-offs. We follow the general principles of [17], where DNA computation is based on strand displace-

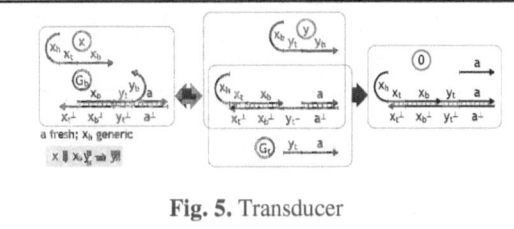

Fig. 5. Transducer

ment on loop-free structures. Other architectures are possible, like computation with hairpins [18], but have not been fully worked out. The four-segment signal structure in [17] yields a full implementation of the combinatorial strand algebra (not shown, but largely implied by that paper). Here we use a novel, simpler, signal structure.

We represent a signal x as a DNA *signal strand* with three segments x_h, x_t, x_b (Figure 1): $x_h = history$, $x_t = toehold$, $x_b = binding$. A toehold is a segment that can reversibly interact with a gate: the interaction can then propagate to the adjacent binding segment. The history is accumulated during previous interactions (it might even be hybridized) and is not part of signal identity. That is, x denotes the equivalence class of signal strands with any history, and a gate is a structure that operates uniformly on such equivalence classes. We generally use arbitrary letters to indicate DNA *segments* (which are single-stranded sequences of bases).

A strand like b,c,d has a *Watson-Crick complement* $(b,c,d)^\perp = d^\perp, c^\perp, b^\perp$ that, as in Figure 2, can partially hybridize with a,b,c along the complementary segments. For two signals x,y, if $x \neq y$ then neither x and y nor x and $y^\perp$ are supposed to hybridize, and this is ensured by appropriate DNA coding of the segments [9],[10]. We assume that all signals are made of 'positive' strands, with 'negative' strands occurring only in gates, and in particular in their input segments; this separation enables the use of 3-letter codes, that helps design independent sequences [10],[20].

The basic computational step of *strand displacement* [17] is shown in Figure 3 for matching single and double strands. This reaction starts with the reversible hybridization of the toehold t with the complementary $t^\perp$ of a

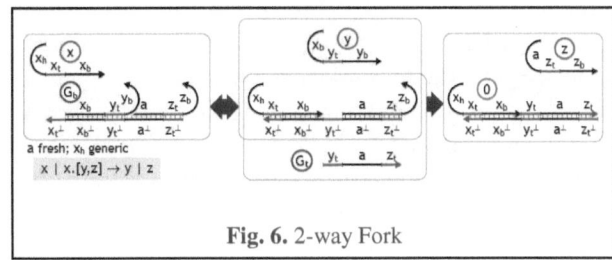

Fig. 6. 2-way Fork

structure that is otherwise double-stranded. The hybridization
can then extend to the binding
segment b by a neutral series
of reactions between base pairs
(*branch migration* [19]) each
going randomly left or right
through small exergy hills, and
eventually ejecting the b strand
when the branch migration
randomly reaches the right
end. The free b strand can in
principle reattach to the dou-

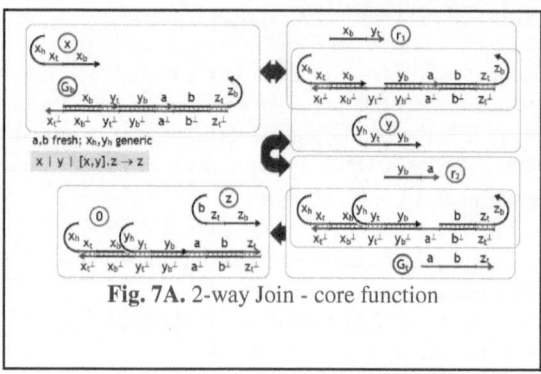

Fig. 7A. 2-way Join - core function

ble-stranded structure, but it has no toehold to do so easily, so the last step is considered irreversible. The simple-minded interpretation of strand displacement is then that
the strand a,b is removed, and the strand b is released irreversibly. The double-
stranded structure is consumed during this process, leaving an inert residual (defined
as one containing no single-stranded toeholds).

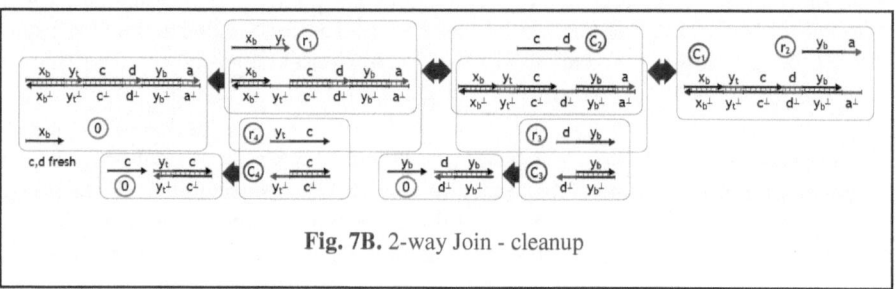

Fig. 7B. 2-way Join - cleanup

Figure 4 shows
the same structure,
but seen as a gate G
absorbing a signal x
and producing noth-
ing (0). The annota-
tion 'x_h generic'
means that the gate
works for all input

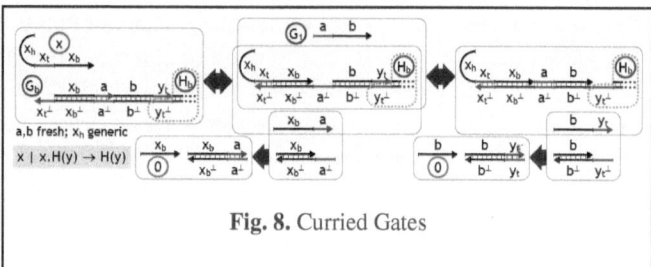

Fig. 8. Curried Gates

histories x_h, as it should. In Figure 5 we implement a gate x.y that transduces a signal x
into a signal y. The gate is made of two separate structures G_b (gate backbone) and G_t
(gate trigger). The forward G_b reaction can cause y to detach because the binding of a
toehold (y_t) is reversible. That whole G_b reaction is reversible via strand displacement
from right to left, but the G_t reaction eventually 'locks' the gate in the state where x is
consumed and y is produced. The annotation 'a fresh' means that the segment 'a' is not
shared by any other gate in the system to prevent interference (while of course the gate
is implemented as a population of identical copies that share that segment). In general,

we take all gate segments to be fresh unless they are non-history segments of input or output signals. Abstractly, an x to y transduction is seen as a single step but the implementation of x.y takes at least two steps, and hence has a different kinetics. This is a common issue in DNA encodings, but its impact can be minimized [17], e.g. in this case by using a large G_t population. In Figure 6 (*cf.* Figure 2 in [17]), we generalize the transducer to a 2-way fork gate, x.[y,z], producing two output signals; this can be extended to n-way fork, via longer trigger strands.

Many designs have been investigated for join gates [5]. The solution shown in Figure 7 admits the coexistence of joins with the same inputs, [x,y].z | [x,y].z', without disruptive crosstalk or preprocessing of the system (not all join gates have this property). It is crucial for join to fire when both its inputs are available, but not to absorb a first input while waiting for the second input, because the second input may never come, and the first input may be needed by another gate (e.g., another join with a third input). The solution is to *reversibly* bind the first input, taking advantage of chemical reversibility. Given two inputs x,y, a 'reversible-AND' G_b backbone releases two strands r_1, r_2, with r_1 providing reversibility while waiting for y (*cf.* Figure 3 in [17]); the trigger G_t finally irreversibly releases the output z (or outputs). In a cleanup phase (Figure 7B), off the critical path, we use a similar reversible-AND C_1 structure (working from right to left) to remove r_1 and r_2 from the system, so that they do not accumulate to slow down further join operations. This phase is initiated by the release of r_2, so we know by construction that both r_1 and r_2 are available. Therefore, the r_3 and r_4 reversibility strands released by C_1 can be cleaned up immediately by C_3, C_4, ending a possible infinite regression of reversible-ANDs. (Note that without the extra c,d segments, a strand $y_t, y_b = y$ would be released.) This gate structure can be easily generalized to n-way join gates by cascading more inputs on the G_b backbone. Alternatively, we can implement a 3-way join from 2-way joins and an extra signal x_0, but this encoding 'costs' a population: $[x_1, x_2, x_3].x_4 \overset{\text{def}}{=} ([x_1, x_2].x_0 \mid x_0.[x_1, x_2])* \mid [x_0, x_3].x_4$.

This completes the implementation of strand algebra in DNA. For the purposes of the next section, however, it is useful to consider also *curried gates* (gates that produce gates). Figure 8 shows a gate x.H(y) that accepts a signal x and activates the backbone H_b of a gate H(y), where H(y) can be any gate with initial toehold $y_t^{\perp}$, including another curried gate. For example, if H(y) is a transducer y.z as in Figure 5, we obtain a curried gate x.y.z such that x | x.y.z → y.z. (The extra a,b segments prevent the release of a strand x_b, y_t that would interfere with r_1 of [x,y].z; see Figure 7A.) This implies that there is an extension of strand algebra with gates of the form G ::= $[x_1,..,x_n].[x'_1,..,x'_m]$ ⋮ $[x_1,..,x_n].G$; this extension can be translated back to the basic strand algebra, e.g. by setting x.y.z = x.w | [w,y].z for a fresh w, but a direct implementation of curried gates is also available.

4 Stochastic Strand Algebra

Stochastic strand algebra is obtained by assigning stochastic rates to gates, and by dropping the unbounded populations, P*. Since the binding strengths of toeholds of the same length are comparable [18], we assume that all gates with the same number n of inputs have the same stochastic rate g_n, collapsing all the gate parameters into a single effective parameter. Although gate rates are fixed, we can vary population sizes in order to achieve desired macroscopic rates. Moreover, as we describe below, it is possible to maintain stable population sizes, and hence to achieve desired stable rate ratios.

In this section $[x_1,..,x_n].[y_1,...,y_m]$ is a stochastic gate of rate g_n, and we write P^k for k parallel copies of P. In a global system state P, the *propensity* of a gate reaction is $(P \ choose \ (x_1 \ | .. \ | \ x_n \ | \ [x_1,..,x_n].[y_1,...,y_m]))\times g_n$; that is, the gate rate g_n multiplied by the number of ways of choosing out of P a multiset consisting of a gate and its n inputs. For example, if $P = x^n \ | \ y^m \ | \ ([x,y].z)^p$ with $x \neq y$, then the propensity of the first reaction in P is $n \times m \times p \times g_2$. A *global transition* from a global state P to a next global state, labeled with its propensity, has then the following form, where \ is multiset difference:

$$P \ \to^{(P \ choose \ (x_1 \ | .. \ | \ x_n \ | \ [x_1,...,x_n].[y_1,...,y_m]))\times g_n} \ P\backslash(x_1 \ | .. \ | \ x_n \ | \ [x_1,..,x_n].[y_1,..,y_m]) \ | \ y_1 \ | \ ... \ | \ y_m$$

The collection of all global transitions from P and from its successive global states forms a labeled transition graph, from which one can extract the Continuous Time Markov Chain of the system [4]. We shall soon use also a curried gate of the form x.G, whose DNA structure is discussed in Section 3, and whose global transitions are:

$$P \ \to^{(P \ choose \ (x \ | \ x.G))\times g_1} \ P\backslash(x \ | \ x.G) \ | \ G$$

In a stochastic system, an unbounded population like P^* has little meaning because its rates are unbounded as well. In stochastic strand algebra we simply drop the P^* construct. In doing so, however, we eliminate the main mechanism for iteration and recursion, and we need to find an alternative mechanism. Rather than P^*, we should instead consider finite populations P^k exerting a stochastic pressure given by the size k. It is also interesting to consider finite populations that *remain* at constant size k: let's indicate them by $P^{=k}$. In particular, $P^{=1}$ represents a single catalyst molecule.

We now show that we can model populations of constant size k by using a bigger buffer population to keep a smaller population at a constant level. Take, e.g., $P = [x,y].z$, and define:

$$P^{=k} \ \overset{\text{def}}{=} \ ([x,y].[z,X])^k \ | \ (X.[x,y].[z,X])^{f(k)} \qquad \text{for a fresh (otherwise unused) signal X}$$

Here $f(k)$ is the size of a large-enough buffer population. A global transition of $P^{=k}$ in context Q (with Q not containing other copies of those gates) is $(Q \ | \ P^{=k}) \to^{((Q \ | \ P^{=k})}$ $choose \ (x|y|[x,y].[z,X]))\times g_2 \ (Q\backslash(x \ | \ y) \ | \ ([x,y].[z,X])^{k-1} \ | \ z \ | \ X \ | \ (X.[x,y].[z,X])^{f(k)})$. For a large enough $f(k)$, the propensity of a next reaction on gate X.[x,y].[z,X] can be made arbitrarily large, so that the two global transitions combined approximate $(Q \ | \ P^{=k}) \to^{((Q \ |}$ $P^{=k}) \ choose \ (x|y|[x,y].[z,X]))\times g_2 \ (Q\backslash(x \ | \ y) \ | \ ([x,y].[z,X])^k \ | \ z \ | \ (X.[x,y].[z,X])^{f(k)-1})$, where the gate population is restored at level k, and the buffer population decreases by 1. We have shown that the reaction propensity in $(Q \ | \ P^{=k})$ can be made arbitrarily close to the reaction propensity in $(Q \ | \ P^k)$, but with the gate population being restored to size k. Moreover, it is possible to periodically replenish the buffer by external intervention *without disturbing the system* (except for the arbitrarily fast reaction speed on X). This provides a practical way of implementing recursion and unbounded computation, by 'topping-up' the buffer populations, without a notion of unbounded population. The construction of a stable population $([x,y].z)^{=k}$ can be carried out also without curried gates, but it then requires balancing the rate of a ternary gate against the desired rate of a binary gate.

We should note that the stochastic strand algebra is a convenient abstraction, but the correspondence with the DNA semantics of Section 3 is not direct. More precisely, it is possible to formulate a formal translation from the stochastic strand algebra to a chemical algebra, by following the figures of Section 3 (considering strand displacement as a single reaction). Such a chemical semantics does not exactly match the global transition semantics given above, because for example a single reaction x | x.y → y is modeled by two chemical reactions. It is possible to define a chemical semantics that approximates the global transition semantics, by using the techniques discussed in [17], but this topic requires more attention that we can provide here.

5   Compiling to Strand Algebra

We give examples of translating other formal languages to strand algebra, in particular translating interacting automata. The interesting point is that by these translations we can map all those formal languages to DNA, by the methods in Section 3.

Finite Stochastic Reaction Networks
We summarize the idea of [17], which shows how to encode with approximate dynamics a stochastic chemical system as a set of DNA signals and gates. A unary reaction $A \to C_1+..+C_n$ is represented as $(A.[C_1,..,C_n])*$. A binary reaction $A+B \to C_1+..+C_n$ is represented as $([A,B].[C_1,..,C_n])*$. The initial solution, e.g. A+A+B, is represented as A | A | B and composed with the populations representing the reactions. For stochastic chemistry, one must replace the unbounded populations with large but finite populations whose sizes and rates are calibrated to provide the desired chemical rates. Because of technical constraints on realizing the rates, one may have to preprocess the system of reactions [17].

Petri Nets
Consider a place-transition Petri Net [13] with places x_i; then, a transition with incoming arcs from places $x_1..x_n$ and outgoing arcs to places $x'_1..x'_m$ is represented as $([x_1,..,x_n].[x'_1..x'_m])*$. The initial marking $\{x_1, .., x_k\}$ is represented as $x_1 | .. | x_k$. The idea is similar to the translation of chemical networks: those can be represented as (stochastic) Petri nets. Conversely (thanks to Cosimo Laneve for pointing this out), a signal can be represented as a marked place in a Petri net, and a gate $[x_1,..,x_n].[x'_1..x'_m]$ as a transition with an additional marked 'trigger' place on the input that makes it fire only once; then, P* can be represented by connecting the transitions of P to refresh the trigger places. Therefore, strand algebra is equivalent to Petri nets. Still, the algebra provides a compositional language for describing such nets, where the gates/transitions are consumed resources.

Finite State Automata

We assume a single copy of the FSA and of the input string. An FSA state is represented as a signal X. The transition matrix is represented as a set of terms $([X,x].[X',\tau])^*$ in parallel, where X is the current state, x is from the input alphabet, X' is the next state, and τ is a signal used to synchronize with the input. For nondeterministic transitions there will be multiple occurrences of the same X and x. The initial state $X_0 \mid \tau$ is placed in parallel with those terms. An input string $x_1,x_2,x_3...$ is encoded as $\tau.[x_1,y_1] \mid [y_1,\tau].[x_2,y_2] \mid [y_2,\tau].[x_3,y_3] \mid ...$ for fresh $y_1,y_2,y_3...$.

Interacting Automata

Interacting automata [4] (a stochastic subset of CCS [11]) are finite state automata that interact with each other over synchronous stochastic channels. An interaction can happen when two automata choose the same channel c_r, with rate r, one as input ($?c_r$) and the other as output ($!c_r$). Intuitively, these automata 'collide' pairwise on complementary exposed surfaces (channels) and change states as a result of the collision. Figure 9 shows two such automata, where each diagram represents a population of identical automata interacting with each other and with other populations (see [3] for many examples). Interacting automata can be faithfully emulated in stochastic strand algebra by generating a binary join gate for each possible collision, and by choosing stable population sizes that produce the prescribed rates. The translation can cause an n^2 expansion of the representation [4].

A system of interacting automata is given by a system E of *equations* of the form X = M, where X is a *species* (an automaton state) and M is a *molecule* of the form $\pi_1;P_1 \oplus ... \oplus \pi_n;P_n$, where $\oplus$ is stochastic choice among possible interactions, P_i are multisets of resulting species, and π_i are either delays τ_r, inputs $?c_r$, or outputs $!c_r$ on a channel c at rate r. For example, in an E_1 population, an automaton in state A_1 can collide by $!a_r$ with an automaton in state B_1 by $?a_r$, resulting in two automata in state A_1:

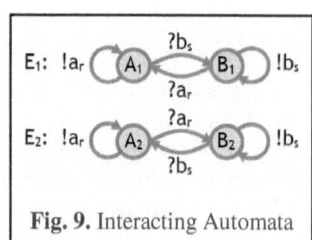

Fig. 9. Interacting Automata

$$E_1: A_1 = !a_r.A_1 \oplus ?b_s.B_1 \qquad E_2: A_2 = !a_r.A_2 \oplus ?a_r.B_2$$

$$B_1 = !b_s.B_1 \oplus ?a_r.A_1 \qquad B_2 = !b_s.B_2 \oplus ?b_s.A_2$$

With initial conditions $A_i^n \mid B_i^m$ (that is, n automata in state A_i and m in state B_i), the Continuous Time Markov Chain semantics of [4] prescribes the *propensities* of the interactions. On channel a_r, in E_1 the propensity is $n \times m \times r$, while in E_2, with two symmetric $?/!$ ways for A_2 to collide with A_2, the propensity is $2 \times (n \text{ choose } 2) \times r = n \times (n-1) \times r$:

$$A_1^n \mid B_1^m: (a_r)\ A_1^n \mid B_1^m \rightarrow^{n \times m \times r} A_1^{n+1} \mid B_1^{m-1} \qquad A_2^n \mid B_2^m: (a_r)\ A_2^n \mid B_2^m \rightarrow^{n \times (n-1) \times r} A_2^{n-1} \mid B_2^{m+1}$$

$$(b_s)\ A_1^n \mid B_1^m \rightarrow^{n \times m \times s} A_1^{n-1} \mid B_1^{m+1} \qquad (b_s)\ A_2^n \mid B_2^m \rightarrow^{m \times (m-1) \times s} A_2^{n+1} \mid B_2^{m-1}$$

Subsequent transitions are computed in the same way. One can also mix E_1,E_2 populations.

The translation of interacting automata to strand algebra is as follows. $E.X.i$ denotes the i-th summand of the molecule associated to X in E; 《 ... 》 and ∪ denote multisets and multiset union to correctly account for multiplicity of interactions; and *Parallel*(S) is the parallel composition of the elements of multiset S. *Strand*(E) is then the translation of a system of equations E, using the stable buffered populations $P^{=k}$ described in Section 4, where g_i are the gate rates of i-ary gates (we assume for simplicity that the round-off errors in r/g_i are not significant and that $r/g_i \geq 1$; otherwise one should appropriately scale the rates r of the original system):

$Strand(E) = Parallel(\ (X.[P])^{=r/g_1}\ s.t.\ \exists i.\ E.X.i = \tau_r;P\ 》\ \cup$
$《([X,Y].[P,Q])^{=r/g_2}\ s.t.\ X{\neq}Y\ and\ \exists i,j,c.\ E.X.i = ?c_r;P\ and\ E.Y.j = !c_r;Q\ 》\ \cup$
$《([X,X].[P,Q])^{=2r/g_2}\ s.t.\ \exists i,j,c.\ E.X.i = ?c_r;P\ and\ E.X.j = !c_r;Q\ 》\ )$

The E_1, E_2 examples above, in particular, translate as follows:

$P_1 = Strand(E_1) = ([B_1,A_1].[A_1,A_1])^{=r/g_2}\ |$ $P_2 = Strand(E_2) = ([A_2,A_2].[B_2,A_2])^{=2r/g_2}\ |$
$\qquad\qquad\qquad ([A_1,B_1].[B_1,B_1])^{=s/g_2}$ $\qquad\qquad\qquad ([B_2,B_2].[A_2,B_2])^{=2s/g_2}$

Initial automata states are translated identically into initial signals and placed in parallel. As described in Section 4, a strand algebra transition from global state $A^n\ |\ B^m\ |\ ([A,B].[C,D])^{=p}$ has propensity $n{\times}m{\times}p{\times}g_2$, and from $A^n\ |\ ([A,A].[C,D])^{=p}$ has propensity $(n\ choose\ 2){\times}p{\times}g_2$. From the same initial conditions $A^n\ |\ B^m$ as in the automata, we then obtain the global strand algebra transitions:

$A_1{}^n|B_1{}^m|P_1 \rightarrow^{n{\times}m{\times}r/g_2{\times}g_2} A_1{}^{n+1}|B_1{}^{m-1}|P'_1$ $A_2{}^n|B_2{}^m|P_2 \rightarrow^{(n{\times}(n-1))/2{\times}2r/g_2{\times}g_2} A_2{}^{n-1}|B_2{}^{m+1}|P'_2$

$A_1{}^n|B_1{}^m|P_1 \rightarrow^{n{\times}m{\times}s/g_2{\times}g_2} A_1{}^{n-1}|B_1{}^{m+1}|P''_1$ $A_2{}^n|B_2{}^m|P_2 \rightarrow^{(m{\times}(m-1))/2{\times}2s/g_2{\times}g_2} A_2{}^{n+1}|B_2{}^{m-1}|P''_2$

which have the same propensities as the interacting automata transitions. Here P'_i, P''_i are systems where a buffer has lost one element, but where the active gate populations that drive the transitions remain at the same level as in P_i. We have shown that the stochastic behavior of interacting automata is preserved by their translation to strand algebra, assuming that the buffers are not depleted.

Figure 10 shows another example: a 3-state automaton and a Gillespie simulation of 1500 such automata with r=1.0. The equation system and its translation to strand algebra are (take, e.g., $r=g_2=1.0$):

$A = !a_r.A \oplus ?b_r.B$ $([A,B].[B,B])^{=r/g_2}\ |$
$B = !b_r.B \oplus ?c_r.C$ $([B,C].[C,C])^{=r/g_2}\ |$
$C = !c_r.C \oplus ?a_r.A$ $([C,A].[A,A])^{=r/g_2}\ |$
$A^{900}\ |\ B^{500}\ |\ C^{100}$ $A^{900}\ |\ B^{500}\ |\ C^{100}$

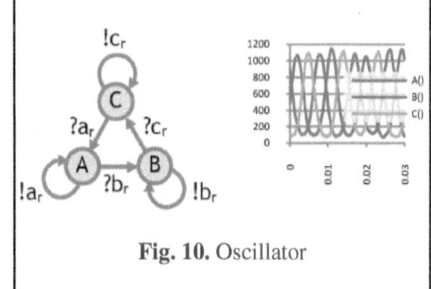

Fig. 10. Oscillator

6 Nested Strand Algebra

The purpose of this section is to allow nesting of join/fork operators in strand algebra, so that natural compound expressions can be written. We provide a uniform translation of this extended language back to $\mathcal{P}$, as a paradigm for the compilation of high(er) level languages to DNA strands. Consider a simple cascade of operations, $?x_1.!x_2.?x_3$, with the meaning of first taking an input ('?') x_1, then producing an output ('!') x_2, and then taking an input x_3. This can be encoded as follows:

$$?x_1.!x_2.?x_3 \quad \overset{\text{def}}{=} \quad x_1.[x_2,x_0] \mid [x_0,x_3].[]$$

where the right hand side is a set of $\mathcal{P}$ combinators, and where x_0 can be chosen fresh so that it does not interfere with other structures (although it will be used by all copies of $?x_1.!x_2.?x_3$).

The nested algebra n$\mathcal{P}$ admits such nesting of operators in general. The main change from the combinatorial $\mathcal{P}$ algebra consists in allowing syntactic nesting after an input or output prefix. This has the consequence that populations can now be nested as well, as in $?x.(P*)$. The new syntax is:

$$P ::= x \;\vdots\; ?[x_1,..,x_n].P \;\vdots\; ![x_1,..,x_n].P \;\vdots\; 0 \;\vdots\; P_1 \mid P_2 \;\vdots\; P* \qquad n \geq 1$$

Here $![x_1,..,x_n].P$ spontaneously releases $x_1,..,x_n$ into the solution and continues as P, while $?[x_1,..,x_n].P$ extracts $x_1,..,x_n$ from the solution (if they are all available) and continues as P. The mixing relation is the same as in $\mathcal{P}$. The reaction relation is modified only in the gate rule:

$?[x_1,..,x_n].P \mid x_1 \mid .. \mid x_n \rightarrow P$	input gate	(e.g.: $?x.0 \mid x \rightarrow 0$)
$![x_1,..,x_n].P \rightarrow x_1 \mid .. \mid x_n \mid P$	output gate	(e.g.: $!x.0 \rightarrow x \mid 0$)

We now show how to compile n$\mathcal{P}$ to $\mathcal{P}$. Let be an infinite lists of distinct signals, and $\mathfrak{F}$ be the set of such $\mathcal{X}$'s. Let $\mathcal{X}_i$ be the i-th signal in the list, $\mathcal{X}_{\geq i}$ be the list starting at the i-th position of $\mathcal{X}$, $evn(\mathcal{X})$ be the even elements of $\mathcal{X}$, and $odd(\mathcal{X})$ be the odd elements. Let $\mathfrak{F}_P$ be the set of those $\mathcal{X} \in \mathfrak{F}$ that do not contain any signal that occurs in P. The unnest algorithm $U(P)\mathcal{X}$, for $P \in$ n$\mathcal{P}$ and $\mathcal{X} \in \mathfrak{F}_P$, is shown in Table 6.1–5. The inner loop $U(X,P)\mathcal{X}$ uses X as the trigger for the translation of P.

6.1.1 Unnest Algorithm

$U(P)\mathcal{X}$	$\overset{\text{def}}{=} \mathcal{X}_0 \mid U(\mathcal{X}_0,P)\mathcal{X}_{\geq 1}$
$U(X, x)\mathcal{X}$	$\overset{\text{def}}{=} X.x$
$U(X, ?[x_1,..,x_n].P)\mathcal{X}$	$\overset{\text{def}}{=} [X,x_1,..,x_n].\mathcal{X}_0 \mid U(\mathcal{X}_0,P)\mathcal{X}_{\geq 1}$
$U(X, ![x_1,..,x_n].P)\mathcal{X}$	$\overset{\text{def}}{=} X.[x_1,..,x_n,\mathcal{X}_0] \mid U(\mathcal{X}_0,P)\mathcal{X}_{\geq 1}$
$U(X, 0)\mathcal{X}$	$\overset{\text{def}}{=} X.[]$
$U(X, P' \mid P'')\mathcal{X}$	$\overset{\text{def}}{=} X.[\mathcal{X}_0,\mathcal{X}_1] \mid U(\mathcal{X}_0,P')_{evn(\mathcal{X}_{\geq 2})} \mid U(\mathcal{X}_1,P'')_{odd(\mathcal{X}_{\geq 2})}$
$U(X, P*)\mathcal{X}$	$\overset{\text{def}}{=} (X.[\mathcal{X}_0,X] \mid U(\mathcal{X}_0,P)\mathcal{X}_{\geq 1})*$

For example, the translations for $?x_1.![x_2,x_3].?x_4.0$ and $?x_1.(x_2{*})$ are:

$$U(?x_1.![x_2,x_3].?x_4.0)\mathcal{X} = {}_0 | [\mathcal{X}_0,x_1].\mathcal{X}_1 | \mathcal{X}_1.[x_2,x_3,\mathcal{X}_2] | [\mathcal{X}_2,x_4].\mathcal{X}_3 | \mathcal{X}_3.[]$$
$$U(?x_1.(x_2{*}))\mathcal{X} = \mathcal{X}_0 | [\mathcal{X}_0,x_1].\mathcal{X}_1 | (\mathcal{X}_1.[\mathcal{X}_2,\mathcal{X}_1] | \mathcal{X}_2.x_2){*}$$

In $?x_1.(x_2{*})$, activating x_1 once causes a linear production of copies of x_2. For an exponential growth of the population one should change $U(X,P{*})\mathcal{X}$ to produce $(X.[\mathcal{X}_0,X,X] | U(\mathcal{X}_0,P')\mathcal{X}_{\geq 1}){*}$. In the nested algebra we can also easily solve systems of recursive definitions; for example: '$X = (?x_1.X | !x_2.Y)$ and $Y = ?x_3.(X | Y)$' can be written as: '$(?X.(?x_1.X | !x_2.Y)){*} | (?Y.?x_3.(X | Y)){*}$'.

As an example, consider a coffee vending machine controller, Vend, that accepts two coins for coffee. An ok is given after the first coin and then either a second coin (for coffee) or an abort (for refund) is accepted:

Vend = ?coin. ![ok,mutex]. (Coffee | Refund)
Coffee = ?[mutex,coin]. !coffee. (Coffee | Vend)
Refund = ?[mutex,abort]. !refund. (Refund | Vend)

Each Vend iteration spawns two branches, Coffee and Refund, waiting for either coin or abort. The branch not taken in the mutual exclusion is left behind; this could skew the system towards one population of branches. Therefore, when the Coffee branch is chosen and the system is reset to Vend, we also spawn another Coffee branch to dynamically balance the Refund branch that was not chosen; conversely for Refund.

7 Contributions and Conclusions

We have introduced strand algebra, a formal language based on a simple relational semantics that is equivalent to place-transition Petri nets (in the current formulation), but allows for compositional descriptions where each component maps directly to DNA structures. Strand algebra connects a simple but powerful class of DNA system to a rich set of techniques from process algebra for studying concurrent systems. Within this framework, it is easy to add operators for new DNA structures, or to map existing operators to alternative DNA implementations. We show how to use strand algebra as an intermediate compilation language, by giving a translation from a more convenient syntax. We also describe a stochastic variant, and a technique for maintaining stable buffered populations to support indefinite and unperturbed computation.

Using strand algebra as a stepping stone, we describe a DNA implementation of interacting automata that preserves stochastic behavior. Interacting automata are one of the simplest process algebras in the literature. Hopefully, more advanced process algebra operators will eventually be implemented as DNA structures, and conversely more complex DNA structures will be captured at the algebraic level, leading to more expressive concurrent languages for programming molecular systems.

I would like to acknowledge the Molecular Programming groups at Caltech for invaluable discussions and corrections. In particular, join and curried gate designs were extensively discussed with Lulu Qian, David Soloveichik and Erik Winfree.

References

1. Benenson, Y., Paz-Elizur, T., Adar, R., Keinan, E., Livneh, Z., Shapiro, E.: Programmable and Autonomous Computing Machine made of Biomolecules. Nature 414(22) (November 2001)
2. Berry, G., Boudol, G.: The Chemical Abstract Machine. In: Proc. 17th POPL, pp. 81–94. ACM, New York (1989)
3. Cardelli, L.: Artificial Biochemistry. In: Condon, A., Harel, D., Kok, J.N., Salomaa, A., Winfree, E. (eds.) Algorithmic Bioprocesses. Springer, Heidelberg (2009)
4. Cardelli, L.: On Process Rate Semantics. Theoretical Computer Science 391(3), 190–215 (2008)
5. Cardelli, L., Qian, L., Soloveichik, D., Winfree, E.: Personal communications
6. Danos, V., Laneve, C.: Formal molecular biology. Theoretical Computer Science 325(1), 69–110 (2004)
7. Fournet, C., Gonthier, G.: The Join Calculus: a Language for Distributed Mobile Programming. In: Barthe, G., Dybjer, P., Pinto, L., Saraiva, J. (eds.) APPSEM 2000. LNCS, vol. 2395, p. 268. Springer, Heidelberg (2002)
8. Hagiya, M.: Towards Molecular Programming. In: Ciobanu, G., Rozenberg, G. (eds.) Modelling in Molecular Biology. Springer, Heidelberg (2004)
9. Kari, L., Konstantinidis, S., Sosík, P.: On Properties of Bond-free DNA Languages. Theoretical Computer Science 334(1-3), 131–159 (2005)
10. Marathe, A., Condon, A.E., Corn, R.M.: On Combinatorial DNA Word Design. J. Comp. Biology 8(3), 201–219 (2001)
11. Milner, R.: Communicating and Mobile Systems: The π-Calculus. Cambridge University Press, Cambridge (1999)
12. Qian, L., Winfree, E.: A Simple DNA Gate Motif for Synthesizing Large-scale Circuits. In: Proc. 14th International Meeting on DNA Computing (2008)
13. Reisig, W.: Petri Nets: An Introduction. Springer, Heidelberg (1985)
14. Regev, A., Panina, E.M., Silverman, W., Cardelli, L., Shapiro, E.: BioAmbients: An Abstraction for Biological Compartments. Theoretical Computer Science 325(1), 141–167 (2004)
15. Sakamoto, K., Kiga, D., Komiya, K., Gouzu, H., Yokoyama, S., Ikeda, S., Sugiyama, H., Hagiya, M.: State Transitions by Molecules. Biosystems 52, 81–91 (1999)
16. Seelig, G., Soloveichik, D., Zhang, D.Y., Winfree, E.: Enzyme-Free Nucleic Acid Logic Circuits. Science 314(8) (2006)
17. Soloveichik, D., Seelig, G., Winfree, E.: DNA as a Universal Substrate for Chemical Kinetics. DNA14
18. Yin, P., Choi, H.M.T., Calvert, C.R., Pierce, N.A.: Programming Biomolecular Self-assembly Pathways. Nature 451, 318–322 (2008)
19. Yurke, B., Mills Jr., A.P.: Using DNA to Power Nanostructures. Genetic Programming and Evolvable Machines archive 4(2), 111–122 (2003)
20. Zhang, D.Y., Turberfield, A.J., Yurke, B., Winfree, E.: Engineering Entropy-driven Reactions and Networks Catalyzed by DNA. Science 318, 1121–1125 (2007)

A Domain-Specific Language for Programming in the Tile Assembly Model*

David Doty and Matthew J. Patitz

Department of Computer Science, Iowa State University, Ames, IA 50011, USA
{ddoty,mpatitz}@cs.iastate.edu

Abstract. We introduce a domain-specific language (DSL) for creating sets of tile types for simulations of the abstract Tile Assembly Model. The language defines objects known as tile templates, which represent related groups of tiles, and a small number of basic operations on tile templates that help to eliminate the error-prone drudgery of enumerating such tile types manually or with low-level constructs of general-purpose programming languages. The language is implemented as a class library in Python (a so-called *internal DSL*), but is presented independently of Python or object-oriented programming, with emphasis on support for a visual editing tool for creating large sets of complex tile types.

1 Introduction

Erik Winfree [15] introduced the abstract Tile Assembly Model (aTAM) as a simplified mathematical model of molecular self-assembly. In particular, it attempts to model Seeman's efforts [10] to coax DNA double-crossover molecules to self-assemble, programmable through the careful selection of sticky ends protruding from the sides of the double-crossover molecules. The basic component of the aTAM is the *tile type*, which defines (many identical copies of) a square tile that can be translated but not rotated, which has glues on each side of strength 0, 1, or 2 (when the temperature is set to 2), each with labels, so that abutting tiles will bind with the given strength if their glue labels match. In particular, by setting the temperature to 2, we may enforce that the abutting sides of two tiles with strength-1 glues provide two input "signals" which determine the tile types that can attach in a given location. Such enforced cooperation is key to all sophisticated constructions in the aTAM.

1.1 Background

The current practice of the design of tile types for the abstract Tile Assembly Model and related models is not unlike early machine-level programming. Numerous theoretical papers [5,12,9,11] focus on the *tile complexity* of systems, the minimum number of tile types required to assemble certain structures such as

---

* This research was partially supported by NSF grants 0652569 and 0728806.

R. Deaton and A. Suyama (Eds.): DNA 15, LNCS 5877, pp. 25–34, 2009.

squares. A major motivation of such complexity measures is the immense level of laboratory effort presently required to create physical implementations of tiles.

Early electronic computers occupied entire rooms to obtain a fraction of the memory and processing power of today's cheapest mobile telephones. This limitation did not stop algorithm developers from creating algorithms, such as divide-and-conquer sorting and the simplex method, that find their most useful niche when executed on data sets that would not have fit on all the memory existing in the world in 1950. Similarly, we hope and expect that the basic components of self-assembly will one day, through the ingenuity of physical scientists, become cheap and easy to produce, not only in quantity but in variety. The *computational* scientists will then be charged with developing disciplined methods of organizing such components so that systems of high complexity can be designed without overwhelming the engineers producing the design. In this paper, we introduce a preliminary attempt at such a disciplined method of controlling the complexity of designing tile assembly systems in the aTAM.

Simulated tile assembly systems of moderate complexity cannot be produced by hand; one must write a computer program that handles the drudgery of looping over related groups of individual tile types. However, even writing a program to produce tile types directly is error-prone, and more low-level than the ways that we tend to think about tile assembly systems.

Fowler [4] suggests that a *domain-specific language (DSL)* is an appropriate tool to introduce into a programming project when the syntax or expressive capabilities of a general-purpose programming language are awkward or inadequate for certain portions of the project. He distinguishes between an *external DSL*, which is a completely new language designed especially for some task (such as SQL for querying or updating databases), and an *internal DSL*, which is a way of "hijacking" the syntax of an existing general-purpose language to express concepts specific to the domain (e.g. Ruby on Rails for writing web applications). An external DSL may be as simple as an XML configuration file, and an internal DSL may be as simple as a class library. In either case the major objective is to express "commands" in the DSL that more closely model the way one thinks about the domain than the "host" language in which the project is written.

If the syntax and semantics of the DSL are precisely defined, this facilitates the creation of a *semantic editor* (text-based or visual), in which programs in the DSL may be produced using a tool that can visually show semantically meaningful information such as compilation or even logical errors, and can help the programmer directly edit the abstract components of the DSL, instead of editing the source code directly. For example, Intellij IDEA and Eclipse are two programs that help Java programmers to do refactorings such as variable renaming or method inlining, which are at their core operations that act directly on the abstract syntax tree of the Java program rather than on the text constituting the source code of the program. We have kept such a tool in mind as a guide for how to appropriately structure the DSL for designing tile systems.

We structure this paper in such a way as to de-emphasize any dependence of the DSL on Python in particular or even on object-oriented class libraries in general. The DSL provides a high-level way of *thinking* about tile assembly programming, which, like any high-level language or other advance in software engineering, not only automates mundane tasks better left to computers, but also *restricts* the programmer from certain error-prone tasks, in order to better guide the design, following the dictum of Antoine de Saint-Exupery that a design is perfected "not when there is nothing left to add, but nothing left to take away."

1.2 Brief Outline of the DSL

We now briefly outline the design of the DSL. Section 2 provides more detail.

The most fundamental component of designing a tile assembly system manually is the tile type. In our DSL, the most fundamental component is an object known as a *tile template*. A tile template represents a group of tile types (each tile type being an *instance* of the tile template), sharing the same input sides, output sides, and function that transforms input signals into output signals. The two fundamental operations of the DSL are *join* and *add transition*. Both make the assumption that each tile template has well-defined input and output sides. In a join, an input tile template A is connected to an output tile template B in a certain direction $d \in \{N, S, E, W\}$, expressing that an instance t_A of A may have on its output side in direction d an instance t_B of B. This expresses an intention that in the growth of the assembly, t_A will be placed first, then t_B, and they will bind with positive strength, with t_A passing information to t_B. Whereas a join is an operation connecting the output side of a tile template to the input side of another, a transition "connects" input sides to output sides within a single tile template, by specifying how to compute information on the output side as a function of information on the input sides. This is called *adding* a transition rather than *setting*, since the information on the output sides may contain multiple independent output signals, and their computations may be specified independently of one another if convenient.

This notion of independent signals is modeled already in other DSLs for the aTAM [1] and for similar systems such as cellular automata [3]. The join operation, however, makes sense in the aTAM but not in a system such as a cellular automaton, where each cell contains the same transition function(s). The notion of tile templates allows one to break an "algorithm" for assembly into separate "stages", each stage corresponding to a different tile template, with each stage being modularized and decoupled from other stages, except through well-defined and restricted signals passed through a join. There is a rough analogy with lines of code in a program: a single line of code may execute more than once, each time executed with the state of memory being different than the previous. Similarly, many different tile types, with different actual signal values, generated from the same tile template, may be placed during the growth of an assembly. Another difference between our language and that of [1] is that our language appears to be more general; rather than being geared specifically toward the creation of geometric shapes, our language is more low-level, but also more general.[1]

---

[1] An extremely crude analogy would be that the progression *manual tile programming* → *Doty-Patitz* → *Becker* is roughly analogous to *machine code* → *C* → *Logo*.

Additionally, [13] presents a DSL which provides a higher level framework for the modeling of various self-assembling systems, while the focus of our DSL on the aTAM creates a more powerful platform for the development of complex tile assembly systems.

This paper is organized as follows. Section 2 describes the DSL for tile assembly programming in more detail and gives examples of design and use. Section 3 concludes the paper and discusses future work and open theoretical questions. Due to space constraints, we refer the reader to [6],which contains a self-contained introduction to the Tile Assembly Model, for a formalism of the aTAM. A preliminary implementation of the DSL and the visual editor can be found at http://www.cs.iastate.edu/~lnsa.

2   Description of Language

The DSL is written as a class library in the Python programming language. It is designed as a set of classes which encapsulate the logical components needed to design a tile assembly system in the aTAM where the temperature value $\tau = 2$.

The DSL is designed around the principal notion that data moves through an assembly as 'signals' which pass through connected tiles as the information encoded in the input glues, allowing a particular tile type to bind in a location, and then, based on the 'computation' performed by that tile type, as the resultant information encoded in the glues of its output edges. (Of course, tiles in the aTAM are static objects so the computation performed by a given tile type is a simple mapping of one set of input glues to one set of output glues which is hardcoded at the time the tile type is created.) Viewed in this way, signals can be seen to propagate as tiles attach and an assembly forms, and it is these signals which dictate the final shape of the assembly. Using this understanding of tile-based self-assembly, we designed the DSL from the standpoint of building tile assembly systems around the transmission and processing of such signals.

We present a detailed overview of the objects and operations provided by the DSL, then demonstrate an example exhibiting the way in which the DSL is used to design a tile set.

2.1   Client-Side Description

The DSL provides a collection of objects which represent the basic design units and a set of operations which can be performed on them. We describe these objects and operations in this section, without describing the underlying data structures and algorithms implementing them.

The DSL strictly enforces the notion of input and output sides for tile types, meaning that any given side can be designated as receiving signals (an input side), sending signals (an output side), or neither (a blank side).

DSL Objects

Tile system. The highest level object is the *tile system* object, which represents the full specification of a temperature 2 tile assembly system. It contains child objects representing the tile set and seed specification, and provides the functionality to write them out to files in a format readable by the ISU TAS simulator [7] (http://www.cs.iastate.edu/~lnsa/software.html).

Tile. In some cases, especially for tiles contained within seed assemblies, there is no computation being performed. In such situations, it may be easier for the programmer to fully specify the glues and properties of a tile type. The *tile* object can be used to easily create such hardcoded tile types.

Tile template. The principle design unit in the DSL is the tile template. This object represents a collection of tile types which share the following properties:

- They have exactly the same input, output, and blank sides.
- The types of signals received/transmitted on every corresponding input/output side are identical.
- The computation performed to transform the input signals to output signals can be performed by the same function, using the values specific to each instantiated tile type.

Logically, a tile template represents the set of tile types which perform the same computation on different values of the same input signal types.

Tile set template. A *tile set template* contains all of the information needed to generate a tile set. It contains the sets of tile and tile template objects included in the tile set, as well as the logic for performing joins, doing error checking, etc. The tile set template object encapsulates information and operations that require communication between more than one tile template object, such as the *join* operation, whereas a tile template is responsible for operations that require information entirely local to the tile template, such as *add transition*.

Signal. A *signal* is simply the name of a variable and the set of allowable values for it. For example, a signal used to represent a binary digit could be defined by giving it the name *bit* and the allowable values 0 and 1.

Transition. *Transitions* are the objects which provide the computational abilities of tile templates. A transition is defined as a set of input signal names, output signal names, and a function which operates on the input signals to yield output signals. The logic of a function can be specified as a table enumerating all input signals and their corresponding outputs, a Python expression, or a Python function, yielding the full power of a general purpose programming language for performing the computations. An example is shown in Figure 1a.

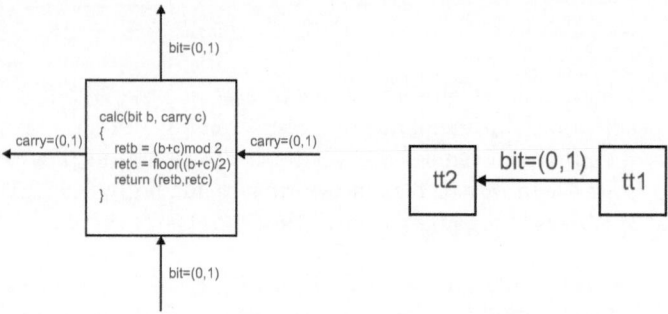

(a) Transition: The function *calc* takes bit/carry values as input from the bottom/right, and computes new bit/carry values which are then output on the top/left.

(b) Join: The tile template on the right (tt1) is sending the signal *bit* with allowable values $(0, 1)$ to the tile template on the left (tt2).

Fig. 1. Logical representations of some DSL objects

DSL Operations

Join. The primary operation between tile templates, which defines the signals that are passed and the paths that they take through an assembly, is called a *join*. Joins are performed between the complementary sides (north and south, or east and west) of two tile templates (which need not be unequal), or a tile and a tile template. If one parameter is a tile, then all signals must be given just one value; otherwise a set of possible values that could be passed between the tile templates is specified (which can still be just one). A join is directional, specifying the direction in which a signal (or set of signals) moves. This direction defines the input and output sides of the tile templates which a join connects. An example is shown in Figure 1b.

Add transition. Transition objects can be added to tile template objects, and indicate how to compute the output signals of a tile template from the input signals. If convenient, a single transition can compute more than one output signal by returning a tuple. Each output signal of a tile template with more than one possible value must have a transition computing it.

Add chooser. There may be multiple tile templates that share the same input signal values on all of their input sides. Depending on the join structure, the library may be able to use annotations (see below) to avoid "collisions" resulting from this, but if the tile templates share joins, then it may not be possible (or desirable) for the library to automatically calculate which tile template should be used to create a tile matching the inputs. In this case, a user-defined function called a *chooser* must be added so that the DSL can ensure that only a single tile type is generated for each combination of input values. This helps to avoid accidental nondeterminism.

Set property. Additional properties such as the display string, tile color, and tile type concentrations can be set using user-defined functions for each property.

2.2 Additional Features

The DSL provides a number of additional, useful features to programmers, a few of which will be described in this section. First, the DSL performs an analysis of the connection structure formed between tile templates by joins and creates *annotations*, or additional information in the form of strings, which are appended to glue labels. These annotations ensure that only tile types created from tile templates which are connected by joins can bind with each other in the directions specified by those joins, preventing the common error of accidentally allowing two tiles to bind that should not because they happen to communicate the same information in the same direction as two unrelated tiles.

Although some forms of 'accidental' nondeterminism are prevented by the DSL, it does provide methods by which a programmer can intentionally create nondeterminism. Specifically, a programmer can either design a chooser function which returns more than one output tile type for a given set of inputs, or one can add *auxiliary inputs* to a tile template, which are input signals that do not come from any direction.

The DSL provides additional error-checking functionality. For example, each tile template must have either one strength-2 input or two strength-1 inputs. As another example, the programmer is forced to specify a chooser function if the DSL cannot automatically determine a unique output tile template, which is a common source of accidental nondeterminism in tile set design.

2.3 Example Construction

In this section, we present an example of how to use the DSL to produce a tile assembly system which assembles a log-width binary counter. In order to slightly simplify this example, the seed row of the assembly, representing the value 1, will be two tiles wide instead of one. All other rows will be of the correct width, i.e. a row representing the value n will be $\lceil \log_2(n + 1) \rceil$ tiles wide.

Figure 2a shows a schematic diagram representing a set of DSL objects, namely tiles, tile templates and joins, which can generate the necessary tile types. It is similar in appearance to how such a diagram appears in the visual editor. The squares labeled *lsbseed* and *msbseed* represent hard-coded tiles for the seed row. The other squares represent tile templates, with the names shown in the middle. The connecting lines represent joins, which each have a direction specified by a terminal arrow and a signal which is named (either *bit* or *carry*) and has a range of allowable values (either 0, or 1, or both). The dashed line between *lsbseed* and *msbseed* represents an implicit join between these two hard-coded tile types because they were manually assigned the same glues.

In addition to the joins, transition functions must be added to the *lsb*, *interior*, and *msb* tile templates to compute their respective output signals, since each has multiple possible values for those signals (whereas *grow*, for instance,

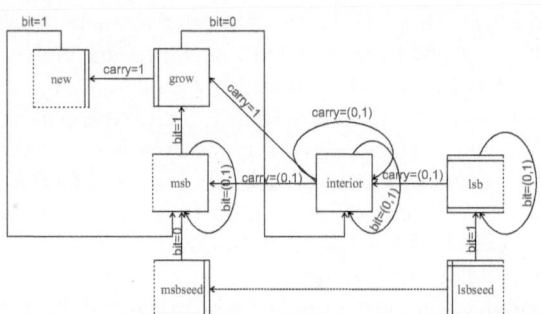

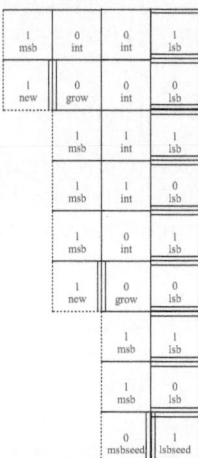

(a) Schematic diagram depicting the tile templates and joins which are used to generate a tile set that self-assembles a log-width binary counter. All east-west joins pass the *carry* signal and south-north joins pass the *bit* signal (although some of them are restricted to subsets of the allowable values $(0,1)$). Transitions are not shown explicitly, but, for example, the tile template *interior* would have a transition as shown in Figure 1a.

(b) First 9 rows of the assembly of a log-width binary counter. Each tile is labeled with its bit value on top and the name of the tile template from which it was generated on the bottom.

Fig. 2. Schematic diagram and partial assembly of generated tile types for log-width counter

which always outputs carry=1 to its west and bit=0 to its north, so needs no transitions). Finally, since the tile templates *msb* and *grow* have overlapping input signal values of bit=1 from *msb* on the south and carry=1 from *interior* on the east, a chooser function must be supplied to indicate that an instance of *grow* should be created for the signals bit=1 and carry=1, and an instance of *msb* should be created in the other three circumstances. Figure 2b shows a partial assembly of tiles generated from the tile template diagram of Figure 2a, without glue labels shown explicitly.

3   Conclusion and Future Work

We have described a domain-specific language (DSL) for creating tile sets in the abstract Tile Assembly Model. This language is currently implemented as a Python class library but is framed as a DSL to emphasize its role as a high-level, disciplined way of thinking about the creation of tile systems.

Semantic Visual Editor. The DSL is implemented as a Python class library, but one thinks of the "real" programming as the creation of visual tile templates and the joins between them, as well as the addition of signal transitions. We have implemented a visual editor that removes the Python programming and allows the direct creation of the DSL objects and execution of its operations. It detects errors, such as a tile template having only one input side of strength 1, while temporarily allowing the editing to continue. Other types of errors, that do not aid the user in being allowed to persist, such as the usage of a single side as both an input and output, are prohibited outright. Many improvements in error detection and handling remain to be implemented and support remains to be added for several features of the DSL.

Avoidance of Accidental Nondeterminism. Initially, our hope had been to design a DSL with the property that it could be used in a straightforward way to design a wide range of existing tile assembly systems, but was sufficiently restricted that it could be guaranteed to produce a deterministic TAS, so long as nondeterminism was not directly and intentionally introduced through chooser functions or auxiliary inputs. Unfortunately, it is straightforward to use the DSL to design a TAS that begins the parallel simulation of two copies of a Turing machine so that if (and only if) the Turing machine halts, two "lines" of tiles are sent growing towards each other to compete nondeterministically, whence the detection of such nondeterminism is undecidable. A common cause of "accidental nondeterminism" in the design of tile assembly systems is the use of a single side of a tile type as both input and output; this sort of error is prevented by the nature of the DSL in assigning each tile template unique, unchanging sets of input sides and output sides. However, we cannot see an elegant way to restrict the DSL further to automatically prevent or statically detect the presence of the sort of "geometric nondeterminism" described in the Turing machine example. The development of such a technique would prove a boon to the designers of tile assembly systems. Conversely, consider Blum's result [2] that any programming language in which all programs are guaranteed to halt requires uncomputably large programs to compute some functions that are trivially computable in a general-purpose language. Perhaps an analogous theorem implies that any restriction statically enforcing determinism also cripples the Tile Assembly Model, so that accidental nondeterminism in tile assembly, like accidental infinite loops in software engineering, would be an unavoidable fact of life, and time is better spent developing formal methods for proving determinism rather than hoping that it could be automatically guaranteed.

Further Abstractions. To better handle very large and complex constructions, it would be useful to support further abstractions, such as grouping sets of tile templates and related joins into "modules" which could then be treated as atomic units. These modules should have well-defined interfaces which allow them to be configured and combined to create more complicated, composite tile sets.

Other Self-Assembly Models. The abstract Tile Assembly Model is a simple but powerful tool for exploring the theoretical capabilities and limitations of molecular self-assembly. However, it is an (intentionally) overly-simplified model. Generalizations of the aTAM, such as the kinetic Tile Assembly Model [14, 16], and alternative models, such as graph self-assembly [8], have been studied theoretically and implemented practically. We hope to leverage the lessons learned from designing the aTAM DSL to guide the design of more advanced DSLs for high-level programming in these alternative models.

Acknowledgements. We thank Scott Summers for help testing and using preliminary versions of the DSL and for helpful comments on this paper.

References

1. Becker, F.: Pictures worth a thousand tiles, a geometrical programming language for self-assembly. Theoretical Computer Science (to appear)
2. Blum, M.: On the size of machines. Information and Control 11(3), 257–265 (1967)
3. Chou, H.-H., Huang, W., Reggia, J.A.: The Trend cellular automata programming environment. SIMULATION: Transactions of The Society for Modeling and Simulation International 78, 59–75 (2002)
4. Fowler, M.: Language workbenches: The killer-app for domain specific languages? (June 2005), http://martinfowler.com/articles/languageWorkbench.html
5. Kao, M.-Y., Schweller, R.T.: Reducing tile complexity for self-assembly through temperature programming. In: Proceedings of the 17th Annual ACM-SIAM Symposium on Discrete Algorithms (SODA 2006), Miami, Florida, January 2006, pp. 571–580 (2007)
6. Lathrop, J.I., Lutz, J.H., Summers, S.M.: Strict self-assembly of discrete Sierpinski triangles. Theoretical Computer Science 410, 384–405 (2009)
7. Patitz, M.J.: Simulation of self-assembly in the abstract tile assembly model with ISU TAS. In: 6th Annual Conference on Foundations of Nanoscience: Self-Assembled Architectures and Devices, Snowbird, Utah, USA, April 20-24 (to appear, 2009)
8. Reif, J.H., Sahu, S., Yin, P.: Complexity of graph self-assembly in accretive systems and self-destructible systems. In: Carbone, A., Pierce, N.A. (eds.) DNA 2005. LNCS, vol. 3892, pp. 257–274. Springer, Heidelberg (2006)
9. Rothemund, P.W.K., Winfree, E.: The program-size complexity of self-assembled squares (extended abstract). In: STOC 2000: Proceedings of the thirty-second annual ACM Symposium on Theory of Computing, pp. 459–468. ACM, New York (2000)
10. Seeman, N.C.: Nucleic-acid junctions and lattices. Journal of Theoretical Biology 99, 237–247 (1982)
11. Soloveichik, D., Winfree, E.: Complexity of compact proofreading for self-assembled patterns. In: Carbone, A., Pierce, N.A. (eds.) DNA 2005. LNCS, vol. 3892, pp. 305–324. Springer, Heidelberg (2006)
12. Soloveichik, D., Winfree, E.: Complexity of self-assembled shapes. SIAM Journal on Computing 36(6), 1544–1569 (2007)
13. Spicher, A., Michel, O., Giavitto, J.-L.: Algorithmic self-assembly by accretion and by carving in MGS. In: Talbi, E.-G., Liardet, P., Collet, P., Lutton, E., Schoenauer, M. (eds.) EA 2005. LNCS, vol. 3871, pp. 189–200. Springer, Heidelberg (2006)
14. Winfree, E.: Simulations of computing by self-assembly. Tech. Report CaltechCSTR:1998.22. California Institute of Technology
15. Winfree, E.: Algorithmic self-assembly of DNA. Ph.D. thesis, California Institute of Technology (June 1998)
16. Winfree, E., Bekbolatov, R.: Proofreading tile sets: Error correction for algorithmic self-assembly. In: Chen, J., Reif, J.H. (eds.) DNA 2003. LNCS, vol. 2943, pp. 126–144. Springer, Heidelberg (2004)

Limitations of Self-assembly
at Temperature One*

David Doty, Matthew J. Patitz, and Scott M. Summers**

Department of Computer Science, Iowa State University, Ames, IA 50011, USA
{ddoty,mpatitz,summers}@cs.iastate.edu

Abstract. We prove that if a set $X \subseteq \mathbb{Z}^2$ weakly self-assembles at temperature 1 in a deterministic (Winfree) tile assembly system satisfying a natural condition known as *pumpability*, then X is a finite union of semi-doubly periodic sets. This shows that only the most simple of infinite shapes and patterns can be constructed using pumpable temperature 1 tile assembly systems, and gives evidence for the thesis that temperature 2 or higher is required to carry out general-purpose computation in a tile assembly system. Finally, we show that general-purpose computation *is* possible at temperature 1 if negative glue strengths are allowed in the tile assembly model.

1 Introduction

Self-assembly is a bottom-up process by which a small number of fundamental components automatically coalesce to form a target structure. In 1998, Winfree [9] introduced the (abstract) Tile Assembly Model (TAM) – an "effectivization" of Wang tiling [7,8] – as an over-simplified mathematical model of the DNA self-assembly pioneered by Seeman [6]. In the TAM, the fundamental components are un-rotatable, but translatable square "tile types" whose sides are labeled with glue "colors" and "strengths." Two tiles that are placed next to each other *interact* if the glue colors on their abutting sides match, and they *bind* if the strength on their abutting sides matches with total strength at least a certain ambient "temperature," usually taken to be 1 or 2.

Despite its deliberate over-simplification, the TAM is an expressive model at temperature 2. The reason is that, at temperature 2, certain tiles are not permitted to bond until *two* tiles are already present to match the glues on the bonding sides, which enables cooperation between different tile types to control the placement of new tiles. Winfree [9] proved that at temperature 2 the TAM is computationally universal and thus can be directed algorithmically.

In contrast, we aim to prove that the lack of cooperation at temperature 1 inhibits the sort of complex behavior possible at temperature 2. Our main

* This research was supported in part by National Science Foundation grants 0652569 and 0728806.
** This author's research was supported in part by NSF-IGERT Training Project in Computational Molecular Biology Grant number DGE-0504304.

R. Deaton and A. Suyama (Eds.): DNA 15, LNCS 5877, pp. 35–44, 2009.

theorem concerns the *weak self-assembly* of subsets of $\mathbb{Z}^2$ using temperature 1 tile assembly systems. Informally, a set $X \subseteq \mathbb{Z}^2$ weakly self-assembles in a tile assembly system $\mathcal{T}$ if some of the tile types of $\mathcal{T}$ are painted black, and $\mathcal{T}$ is guaranteed to self-assemble into an assembly α of tiles such that X is precisely the set of integer lattice points at which α contains black tile types. As an example, Winfree [9] exhibited a temperature 2 tile assembly system, directed by a clever XOR-like algorithm, that weakly self-assembles a well-known set, the discrete Sierpinski triangle, onto the first quadrant. Note that the underlying *shape* (set of all points containing a tile, whether black or not) of Winfree's construction is an infinite canvas that covers the entire first quadrant, onto which a more sophisticated set, the discrete Sierpinski triangle, is painted.

We show that under a plausible assumption, temperature 1 tile systems weakly self-assemble only a limited class of sets. To prove our main result, we require the hypothesis that the tile system is *pumpable*. Informally, this means that every sufficiently long path of tiles in an assembly of this system contains a segment in which the same tile type repeats (a condition clearly implied by the pigeonhole principle), and that furthermore, the subpath between these two occurrences can be repeated indefinitely ("pumped") along the same direction as the first occurrence of the segment, without "colliding" with a previous portion of the path. We give a counterexample in Section 3 (Figure 1) of a path in which the same tile type appears twice, yet the segment between the appearances cannot be pumped without eventually resulting in a collision that prevents additional pumping. The hypothesis of pumpability states (roughly) that in every sufficiently long path, despite the presence of some repeating tiles that cannot be pumped, *there exists* a segment in which the same tile type repeats that *can* be pumped. In the above-mentioned counterexample, the paths constructed to create a blocked segment always contain some previous segment that *is* pumpable. We conjecture that this phenomenon, pumpability, occurs in every temperature 1 tile assembly system that produces a unique infinite structure. We discuss this conjecture in greater detail in Section 5.

A *semi-doubly periodic* set $X \subseteq \mathbb{Z}^2$ is a set of integer lattice points with the property that there are three vectors $\boldsymbol{b}$ (the "base point" of the set), $\boldsymbol{u}$, and $\boldsymbol{v}$ (the two periods of the set), such that $X = \{\ \boldsymbol{b} + n \cdot \boldsymbol{u} + m \cdot \boldsymbol{v} \mid n, m \in \mathbb{N}\ \}$. That is, a semi-doubly periodic set is a set that repeats infinitely along two vectors (linearly independent vectors in the non-degenerate case), starting at some base point $\boldsymbol{b}$. We show that any directed, pumpable, temperature 1 tile assembly system weakly self-assembles a set $X \subseteq \mathbb{Z}^2$ that is a finite union of semi-doubly periodic sets.

It is our contention that weak self-assembly captures the intuitive notion of what it means to "compute" with a tile assembly system. For example, the use of tile assembly systems to build shapes is captured by requiring all tile types to be black, in order to ask what set of integer lattice points contain any tile at all (so-called *strict self-assembly*). However, weak self-assembly is a more general notion. For example, Winfree's above mentioned result shows that the discrete Sierpinski triangle weakly self-assembles at temperature 2 [5], yet this shape does not strictly self-assemble at *any* temperature [2]. Hence weak self-assembly allows for a more relaxed notion of set building, in which intermediate space can be used for computation, without requiring that the space filled to carry out the computation also represent the final result of the computation.

As another example, there is a standard construction [9] by which a single-tape Turing machine may be simulated by a temperature 2 tile assembly system. Regardless of the semantics of the Turing machine (whether it decides a language, enumerates a language, computes a function, etc.), it is routine to represent the result of the computation by the weak self-assembly of some set. For example, Patitz and Summers [3] showed that for any decidable language $A \subseteq \mathbb{N}$, A's projection along the X-axis (the set $\{ (x,0) \in \mathbb{N}^2 \mid x \in A \}$) weakly self-assembles in a temperature 2 tile assembly system.

Our result is motivated by the thesis that if a tile assembly system can reasonably be said to "compute", then the result of this computation can be represented in a straightforward manner as a set $X \subseteq \mathbb{Z}^2$ that weakly self-assembles in the tile assembly system, or a closely related tile assembly system. Our examples above provide evidence for this thesis, although it is as informal and unprovable as the Church-Turing thesis. On the basis of this claim, and the triviality of semi-doubly periodic sets (shown more formally in Observation 1), we conclude that our main result implies that directed, pumpable, temperature 1 tile assembly systems are incapable of general-purpose deterministic computation, without further relaxing the model.

2   The Abstract Tile Assembly Model

We work in the 2-dimensional discrete space $\mathbb{Z}^2$. Define the set U_2 to be the set of all *unit vectors*, i.e., vectors of length 1 in $\mathbb{Z}^2$. We write $[X]^2$ for the set of all 2-element subsets of a set X. All *graphs* here are undirected graphs, i.e., ordered pairs $G = (V, E)$, where V is the set of *vertices* and $E \subseteq [V]^2$ is the set of *edges*.

Our notation is that of [2], which contains a self-contained introduction to the Tile Assembly Model for the reader unfamiliar with the model.

Intuitively, a tile type t is a unit square that can be translated, but not rotated, having a well-defined "side $\boldsymbol{u}$" for each $\boldsymbol{u} \in U_2$. Each side $\boldsymbol{u}$ of t has a "glue" of "color" $\mathrm{col}_t(\boldsymbol{u})$ – a string over some fixed alphabet Σ – and "strength" $\mathrm{str}_t(\boldsymbol{u})$ – a nonnegative integer – specified by its type t. Two tiles t and t' that are placed at the points $\boldsymbol{a}$ and $\boldsymbol{a} + \boldsymbol{u}$ respectively, *bind* with *strength* $\mathrm{str}_t(\boldsymbol{u})$ if and only if $(\mathrm{col}_t(\boldsymbol{u}), \mathrm{str}_t(\boldsymbol{u})) = (\mathrm{col}_{t'}(-\boldsymbol{u}), \mathrm{str}_{t'}(-\boldsymbol{u}))$.

Given a set T of tile types, an *assembly* is a partial function $\alpha : \mathbb{Z}^2 \dashrightarrow T$, with points $\boldsymbol{x} \in \mathbb{Z}^2$ at which $\alpha(\boldsymbol{x})$ is undefined interpreted to be empty space, so that $\mathrm{dom}\,\alpha$ is the set of points with tiles. α is *finite* if $|\mathrm{dom}\,\alpha|$ is finite. For assemblies α and α', we say that α is a *subconfiguration* of α', and write $\alpha \sqsubseteq \alpha'$, if $\mathrm{dom}\,\alpha \subseteq \mathrm{dom}\,\alpha'$ and $\alpha(\boldsymbol{x}) = \alpha'(\boldsymbol{x})$ for all $\boldsymbol{x} \in \mathrm{dom}\,\alpha$.

Let α be an assembly and $B \subseteq \mathbb{Z}^2$. α *restricted to* B, written as $\alpha \upharpoonright B$, is the unique assembly satisfying $(\alpha \upharpoonright B) \sqsubseteq \alpha$, and $\mathrm{dom}\,(\alpha \upharpoonright B) = B$. If π is a sequence over $\mathbb{Z}^2$ (such as a path), then we write $\alpha \upharpoonright \pi$ to mean α restricted to the set of points in π. If $A \subseteq \operatorname{dom} \alpha$, we write $\alpha \setminus A = \alpha \upharpoonright (\operatorname{dom} \alpha - A)$. If $\mathbf{0} \neq \mathbf{v} \in \mathbb{Z}^2$, then the *translation of α by $\mathbf{v}$* is defined as the assembly $(\alpha + \mathbf{v})$ satisfying, for all $\mathbf{a} \in \mathbb{Z}^2$, $(\alpha + \mathbf{v})(\mathbf{a}) = \alpha(\mathbf{a})$ if $\mathbf{a} - \mathbf{v} \in \operatorname{dom} \alpha$, and undefined otherwise.

A *grid graph* is a graph $G = (V, E)$ in which $V \subseteq \mathbb{Z}^2$ and every edge $\{\mathbf{a}, \mathbf{b}\} \in E$ has the property that $\mathbf{a} - \mathbf{b} \in U_2$. The *binding graph of* an assembly α is the grid graph $G_\alpha = (V, E)$, where $V = \operatorname{dom} \alpha$, and $\{\mathbf{m}, \mathbf{n}\} \in E$ if and only if (1) $\mathbf{m} - \mathbf{n} \in U_2$, (2) $\operatorname{col}_{\alpha(\mathbf{m})} (\mathbf{n} - \mathbf{m}) = \operatorname{col}_{\alpha(\mathbf{n})} (\mathbf{m} - \mathbf{n})$, and (3) $\operatorname{str}_{\alpha(\mathbf{m})} (\mathbf{n} - \mathbf{m}) > 0$. An assembly is *$\tau$-stable*, where $\tau \in \mathbb{N}$, if it cannot be broken up into smaller assemblies without breaking bonds of total strength at least τ (i.e., if every cut of G_α cuts edges, the sum of whose strengths is at least τ).

A *tile assembly system* (*TAS*) is an ordered triple $\mathcal{T} = (T, \sigma, \tau)$, where T is a finite set of tile types, σ is a seed assembly with finite domain, and τ is the temperature. In subsequent sections of this paper, we assume that $\tau = 1$ unless explicitly stated otherwise. An *assembly sequence* in a TAS $\mathcal{T} = (T, \sigma, 1)$ is a (possibly infinite) sequence $\boldsymbol{\alpha} = (\alpha_i \mid 0 \leq i < k)$ of assemblies in which $\alpha_0 = \sigma$ and each α_{i+1} is obtained from α_i by the "τ-stable" addition of a single tile. The *result* of an assembly sequence $\boldsymbol{\alpha}$ is the unique assembly $\operatorname{res}(\boldsymbol{\alpha})$ satisfying $\operatorname{dom} \operatorname{res}(\boldsymbol{\alpha}) = \bigcup_{0 \leq i < k} \operatorname{dom} \alpha_i$ and, for each $0 \leq i < k$, $\alpha_i \sqsubseteq \operatorname{res}(\boldsymbol{\alpha})$.

We write $\mathcal{A}[\mathcal{T}]$ for the *set of all producible assemblies of $\mathcal{T}$*. An assembly α is *terminal*, and we write $\alpha \in \mathcal{A}_\square[\mathcal{T}]$, if no tile can be stably added to it. We write $\mathcal{A}_\square[\mathcal{T}]$ for the *set of all terminal assemblies of $\mathcal{T}$*. A TAS $\mathcal{T}$ is *directed*, or *produces a unique assembly*, if it has exactly one terminal assembly i.e., $|\mathcal{A}_\square[\mathcal{T}]| = 1$. The reader is cautioned that the term "directed" has also been used for a different, more specialized notion in self-assembly [1]. We interpret "directed" to mean "deterministic", though there are multiple senses in which a TAS may be deterministic or nondeterministic.

A set $X \subseteq \mathbb{Z}^2$ *weakly self-assembles* if there exists a TAS $\mathcal{T} = (T, \sigma, 1)$ and a set $B \subseteq T$ (B constitutes the "black" tiles) such that $\alpha^{-1}(B) = X$ holds for every assembly $\alpha \in \mathcal{A}_\square[\mathcal{T}]$. A set X *strictly self-assembles* if there is a TAS $\mathcal{T}$ for which every assembly $\alpha \in \mathcal{A}_\square[\mathcal{T}]$ satisfies $\operatorname{dom} \alpha = X$. Note that if X strictly self-assembles, then X weakly self-assembles. (Let all tiles be black.)

3   Pumpability, Finite Closures, and Combs

Throughout this section, let $\mathcal{T} = (T, \sigma, 1)$ be a directed TAS, and α be the unique assembly satisfying $\alpha \in \mathcal{A}_\square[\mathcal{T}]$. Further, we assume without loss of generality that σ consists of a single "seed" tile type placed at the origin.

Given, $\mathbf{0} \neq \mathbf{v} \in \mathbb{Z}^2$, a *$\mathbf{v}$-semi-periodic path in α originating at* $\mathbf{a}_0 \in \operatorname{dom} \alpha$ is an infinite, simple path $\pi = (\mathbf{a}_0, \mathbf{a}_1, \ldots)$ in the binding graph G_α such that there is a constant $k \in \mathbb{N}$ such that, for all $j \in \mathbb{N}$, $\pi[j + k] = \pi[j] + \mathbf{v}$, and $\alpha(\pi[j + k]) = \alpha(\pi[j])$. Intuitively, $\mathbf{v}$ is the "geometric" period of the path – the straight-line vector between two repeating tile types – while k is the "linear" period – the number of tiles that must be traversed along the path before the tile types repeat, which is at least $\|\mathbf{v}\|_1$, but possibly larger if the segment from $\pi[j]$ to $\pi[j] + \mathbf{v}$ is "winding".

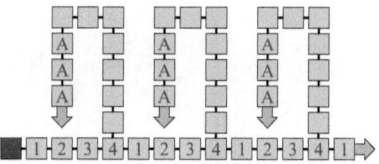

Fig. 1. An assembly containing a path with repeating tiles A-A that do *not* form a pumpable segment, because they are blocked from infinite growth by the existing assembly. Note, however, that any sufficiently long path from the origin (at the left) contains a pumpable segment, namely the repeating segment 1-2-3-4-1 along the bottom row, which can be pumped infinitely to the right.

An *eventually v-semi-periodic path in α originating at $a_0 \in \text{dom } \alpha$* is an infinite, simple path $\pi = (a_0, a_1, \ldots)$ in the binding graph G_α for which there exists $s \in \mathbb{N}$ such that the path $\pi' = (\pi[s], \pi[s+1], \ldots)$ is a v-semi-periodic path in α originating at $\pi[s]$. Let the *initial segment length* be the smallest index s^* such that $\pi' = (\pi[s^* - k], \pi[s^* - k + 1], \ldots)$ is a v-semi-periodic path originating at the point $\pi[s^* - k]$. The *initial segment of π* is the path $\pi[0 \ldots i^* - 1]$ (for technical reasons, we enforce the initial segment of π to contain the simple path $\pi[0 \ldots s]$ along with one period of π'). The *tail of π* is $\pi[s^* \ldots]$. Note that the tail of an eventually v-semi-periodic path is simply a v-semi-periodic path originating at $\pi[s^*]$. An *eventually v-periodic path in α* is an eventually v-periodic path in α originating at a_0 for some $a_0 \in \text{dom } \alpha$. We say that π is a *v-periodic path in α* if $\pi = (\ldots, a_{-1}, a_0, a_1, \ldots)$ is a two-way infinite simple path such that, for all $j \in \mathbb{Z}$, $\alpha\left(\pi[j] + v\right) = \alpha\left(\pi[j]\right)$.

Let $w, x \in \text{dom } \alpha$, π be a simple path from w to x in the binding graph G_α, and $i, k \in \mathbb{N}$ with $0 \le i < k \le |\pi|$. We say that π has a *pumpable segment $\pi[i \ldots k]$* (with respect to $v = \pi[k] - \pi[i]$) if there exists a v-semi-periodic path π' in α originating at $\pi[i]$ and $\pi'[0 \ldots k - i] = \pi[i \ldots k]$.

Intuitively, the path π has a pumpable segment if, after some initial sequence of tile types, it consists of a subsequence of tile types which is repeated in the same direction an infinite number of times, one after another. Figure 1 shows an assembly in which the same tile type repeats along a path, but the segment between the occurrences is not pumpable.

Let $c \in \mathbb{N}$ and $v \in \mathbb{Z}^2$. The *diamond of radius c centered about the point v* is the set of points defined as $D(c, v) = \{ (x, y) + v \mid |x| + |y| \le c \}$. Let $w, x \in \text{dom } \alpha$, $c \in \mathbb{N}$, and π be a simple path from w to x in the binding graph G_α. We say that π is a *pumpable path from w to x in α* if it contains a pumpable segment $\pi[i \ldots k]$ for some $i, k \in \mathbb{N}$ such that $0 \le i < k \le |\pi|$. We say that $\mathcal{T}$ is *c-pumpable* if there exists $c \in \mathbb{N}$ such that for every $w, x \in \text{dom } \alpha$ with $x \notin D(c, w)$, there exists a pumpable path π from x to w in α.

Figure 2 shows, from left to right, (1) a partially complete assembly beginning from the (dark grey) seed tile, where the dark notches between adjacent tiles represent strength 1 bonds, and a tile selected for the example, (2) the full path leading from the seed to the selected tile, (3) the tile types for a segment of the

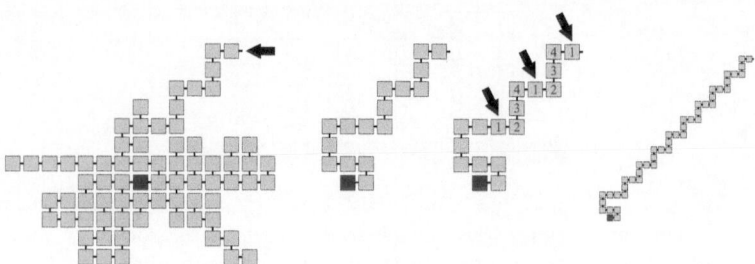

Fig. 2. A partial assembly, selected path, pumpable segment, and pumpable path

path, showing the repeating pattern of tile types '1-2-3-4', and (4) an extended version of the path which shows its ability to be pumped.

In our proof, it is helpful to consider extending an assembly in such a way that no individual tile in the existing assembly is extended by more than a finite amount (though an infinite assembly may have an infinite number of tiles that can each be extended by a finite amount). We call such an extension the *finite closure* of the assembly, and define it formally as follows. Let $\alpha' \in \mathcal{A}[\mathcal{T}]$. We say that the *finite closure* of α' is the unique assembly $\mathcal{F}(\alpha')$ satisfying

1. $\alpha' \sqsubseteq \mathcal{F}(\alpha')$, and
2. dom $\mathcal{F}(\alpha')$ is the set of all points $\boldsymbol{x} \in \mathbb{Z}^2$ such that every infinite simple path in the binding graph G_α containing $\boldsymbol{x}$ intersects dom α'.

Intuitively, this means that if we extend α' by only those "portions" that will eventually stop growing, the finite closure is the super-assembly that will be produced. That is, any attempt to "leave" α' through the finite closure and go infinitely far will eventually run into α' again. If α' is terminal, then α' is its own finite closure. Note that in general, the finite closure of an assembly α' is not the result of adding finitely many tiles to α'. For instance, if infinitely many points of α' allow exactly one tile to be added, the finite closure adds infinitely many points to α'. However, the finite closure of a finite assembly is always a finite assembly.

For an example of a finite closure of an assembly, see Figure 3. Figure 3c shows the terminal assembly which consists of three rows of tiles that continue infinitely

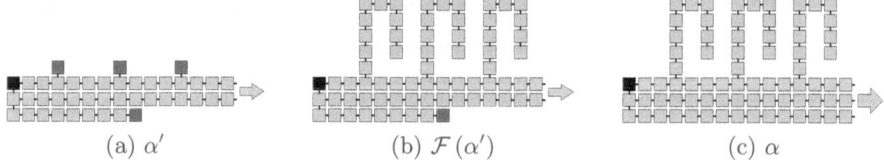

(a) α' \qquad\qquad (b) $\mathcal{F}(\alpha')$ \qquad\qquad (c) α

Fig. 3. Example of a finite closure. The dark gray points represent locations at which tiles can attach.

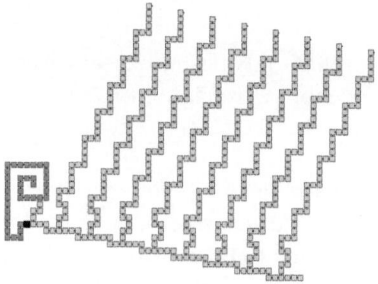

Fig. 4. Example of a comb connected to a hard-coded assembly (dark tiles that spiral). The seed is the tile at the center of the spiral. The comb originates at the bottom most black tile. Note that a finite number of "attempts at teeth" (one in the above example) may be blocked before the infinite teeth are allowed to grow.

to the right (denoted by the arrow), with a 10 tile upward projection occurring at every fourth column. Figure 3a shows an assembly α' which consists of two rows of tiles continuing infinitely to the right but with an incomplete bottom row and none of the full upward projections. Figure 3b shows $\mathcal{F}(\alpha')$, which is the finite closure of α'. Notice that an infinite number of tiles were added to α' to create $\mathcal{F}(\alpha')$ since an infinite number of upward projections were added, each consisting of only 10 tiles. However, also note that the bottom row is not grown because that row consists of an infinite path and therefore cannot be part of the finite closure.

Let $\pi^{\rightarrow}$ be an eventually $\boldsymbol{v}$-semi-periodic path in α originating at $\boldsymbol{0}$ where $\boldsymbol{0} \neq \boldsymbol{v} \in \mathbb{Z}^2$. Let $\boldsymbol{u} \in \mathbb{Z}^2$ such that $\boldsymbol{u} \neq z \cdot \boldsymbol{v}$ for all $z \in \mathbb{R}$. Suppose that there is a point $\boldsymbol{b}$ on the tail of $\pi^{\rightarrow}$ such that there is an eventually $\boldsymbol{u}$-semi-periodic path $\pi^{\uparrow}$ in α originating at $\boldsymbol{b}$ such that $\pi^{\rightarrow} \cap \pi^{\uparrow} = \{\boldsymbol{b}\}$. Define the assembly

$$\alpha^* = \alpha \restriction \left(\pi^{\rightarrow} \cup \bigcup_{n \in \mathbb{N}} \left(\pi^{\uparrow} + n \cdot \boldsymbol{v} \right) \right).$$ It is easy to see that α^* need not be a

producible assembly. We say that α^* is a *comb in* (with respect to $\pi^{\rightarrow}$ and $\pi^{\uparrow}$) if, for every $n \in \mathbb{N}$, $\alpha^* \restriction \pi^{\uparrow} + n \cdot \boldsymbol{v} = \alpha^* \restriction \left(\pi^{\uparrow} + n \cdot \boldsymbol{v} \right)$. We refer to the assembly $\alpha^* \restriction \pi$ as the *base* of α^*. For any $n \in \mathbb{N}$, we define the nth *tooth* of α^* to be the assembly $\alpha \restriction \left(\pi^{\uparrow} + n \cdot \boldsymbol{v} \right)$. We say that the comb α^* *starts at the point* $\pi[s]$ (the point at which the eventually-periodic path $\pi^{\rightarrow}$ becomes periodic). It follows from the definition that, if α^* is a comb in α, then $\alpha^* \in \mathcal{A}[\mathcal{T}]$.

See Figure 4 for an example of a comb. A comb, intuitively, is a generalization of the assembly in which an infinite periodic one-way path (the base) grows along the positive x-axis, and once per period, an infinite periodic path (a tooth) grows in the positive y direction, creating an infinite number of "teeth". The generalizations are that (1) the base and teeth need not run parallel to either axis, and (2), the teeth may have some initial hard-coded tiles before the repeating periodic segment begins. Also, it is possible at temperature 1 to build multiple combs with the same base, but with different teeth, growing in either direction.

4   Main Result

We show in this section that only "simple" sets weakly self-assemble in directed, pumpable tile assembly systems at temperature 1. A set $X \subseteq \mathbb{Z}^2$ is *semi-doubly periodic* if there exist three vectors $\boldsymbol{b}$, $\boldsymbol{u}$, and $\boldsymbol{v}$ such that $X = \{\, \boldsymbol{b} + n \cdot \boldsymbol{u} + m \cdot \boldsymbol{v} \mid n, m \in \mathbb{N} \,\}$.

Note that a set that is periodic along only one dimension is also semi-doubly periodic, by our definition, since this corresponds to the condition that exactly one of the vectors $\boldsymbol{u}$ or $\boldsymbol{v}$ is equal to $\boldsymbol{0}$ (or if $\boldsymbol{u} = \boldsymbol{v}$). Similarly, if $\boldsymbol{u} = \boldsymbol{v} = \boldsymbol{0}$, then the definition of semi-doubly periodic is equivalent to A being a singleton set. The following observation justifies the intuition that finite unions of semi-doubly periodic sets constitute only the computationally simplest subsets of $\mathbb{Z}^2$.

Observation 1. *Let $A \subseteq \mathbb{Z}^2$ be a finite union of semi-doubly periodic sets. Then the unary languages $L_{A,x} = \{\, 0^{|x|} \mid (x, y) \in A \text{ for some } y \in \mathbb{Z} \,\}$ and $L_{A,y} = \{\, 0^{|y|} \mid (x, y) \in A \text{ for some } x \in \mathbb{Z} \,\}$ consisting of the unary representations of the projections of A onto the x-axis and y-axis, respectively, are regular languages.*

The following theorem is the main result of this paper.

Theorem 2. *Let $\mathcal{T} = (T, \sigma, 1)$ be a directed, pumpable TAS. If a set $X \subseteq \mathbb{Z}^2$ weakly self-assembles in $\mathcal{T}$, then X is a finite union of semi-doubly periodic sets.*

The proof idea of Theorem 2 is as follows. Suppose that $\alpha \in \mathcal{A}_\square[\mathcal{T}]$. Note that α is unique since $\mathcal{T}$ is directed. Either α is an infinite "grid" that fills the plane, or there exists finitely many semi-doubly periodic combs and semi-periodic paths that, taken together, "cover" every point in X (in the sense that each such point is in the finite closure of one of these combs or paths).

The reason for this is that each comb is defined by two vectors $\boldsymbol{u}$ (the base) and $\boldsymbol{v}$ (the teeth), and these vectors form a "basis" for the space of points located within the cone formed by the base and the first tooth of the comb. While the vectors do not reach every point in this cone, they reach within a constant distance of every point in the cone, and the doubly periodic regularity of the teeth and base enforces doubly periodic regularity in between the teeth as well. Of course, not all combs have teeth, in which case the comb is just a periodic path.

We associate each black tile with some periodic path or comb that begins in a fixed radius about the origin (utilizing the fact that a path cannot go far from the origin without having pumpable segments that can be used to construct a periodic path or comb). The finite number of combs and periodic paths originating within this radius tells us that the number of semi-doubly periodic sets of (locations tiled by) black tiles that they each define is finite.

A simple application of Theorem 2 is that no discrete self-similar fractal weakly (and hence strictly) self-assembles in any temperature 1 tile assembly system that is pumpable and directed. Since Winfree [9] showed that one particular discrete self-similar fractal, the discrete Sierpinski triangle, weakly self-assembles at temperature 2, this provides a concrete example of computation that is possible (and simple) at temperature 2, but impossible at temperature 1, assuming directedness and pumpability.

5   Conclusion

We have studied the class of shapes that self-assemble in Winfree's abstract tile assembly model at temperature 1. We introduced the notion of a pumpable temperature 1 tile assembly system and then proved that, if X weakly self-assembles in a pumpable, directed tile assembly system, then X is necessarily "simple" in the sense that X is merely a finite union of semi-doubly periodic sets. We conjecture that every directed TAS producing an infinite assembly at temperature 1 is pumpable.   Proving this conjecture would imply that every directed tile set weakly self-assembles a finite union of doubly periodic sets.

We also leave open the question of whether the hypothesis of directedness may be removed. We use the property of directedness at many points in our proof, but in some cases, a more careful and technically convoluted argument could be used to show that the tile set need not be directed. Intuitively, an undirected tile set T that weakly self-assembles a set $X \subseteq \mathbb{Z}^2$ is deterministic in that all terminal assemblies of T "paint" exactly the points in X black, but is nondeterministic in the sense that different terminal assemblies of T may place different tiles in the same location (including different black tiles at locations in X), and may even place non-black tiles at locations in one terminal assembly that are left empty in other terminal assemblies. Undirected tile assembly systems that weakly self-assemble a unique set X exist, but in every case that we know of, the undirected tile set may be replaced by a directed tile set self-assembling the same set.

If both hypotheses of directedness and pumpability could be removed from the entire proof, then our main result would settle the case of computation via self-assembly at temperature 1, by showing that every temperature 1 tile assembly system weakly self-assembles a finite union of semi-doubly periodic sets if it weakly self-assembles any set at all. As indicated in the introduction, we interpret this statement to imply that general-purpose deterministic computation is not possible with temperature 1 tile assembly systems.

Finally, we make the observation that universal computation at temperature 1 *is* possible under variations of the self-assembly model. For instance, a construction personally communicated by Matt Cook (also mentioned briefly in [4]) establishes that the abstract TAM is computationally universal with respect to directed, temperature 1 tile assembly systems that place tiles in three spatial dimensions.   Moreover, universality can also be achieved if negative glue strengths and interaction between differently colored glues (a so-called *non-diagonal strength function*) are allowed.

Theorem 3. *For every single-tape Turing machine M, there is a tile set T with negative, non-diagonal glue strengths, which simulates M in the following way. Given an input string x, define $\mathcal{T}_x = (T, \sigma_x, 1)$ to be the temperature 1 TAS where σ_x is the seed assembly satisfying $\operatorname{dom} \sigma_x = \{0, \dots, |x| - 1\} \times \{0\}$ that encodes the initial configuration of M. Then $\mathcal{T}_x$ simulates the computation of*

$M(x)$, with the configuration of $M(x)$ after n steps represented by the line $y = n$ in the terminal assembly of $\mathcal{T}_x$.

Intuitively, by introducing negative glue strengths, we allow for the cooperation that is impossible with only nonnegative glue strengths. The difference with temperature 2 is that, at temperature 2, one may enforce that no tile appears at a certain position until two neighbors are present. At temperature 1 and with negative glue strengths, on the contrary, we do not enforce that two tiles are present before a position can be given a tile. However, we enforce that if the wrong tile binds in this position, eventually the error is corrected by the presence of a neighboring tile which forces the removal of the incorrect tile using negative glue strengths.

Acknowledgments. We wish to thank Maria Axenovich, Matt Cook, and Jack Lutz for useful discussions, and anonymous referees for corrections. We would especially like to thank Niall Murphy, Turlough Neary, Anthony K. Seda and Damien Woods for inviting us to present a preliminary version of this research at the International Workshop on The Complexity of Simple Programs, University College Cork, Ireland on December 6th and 7th, 2008. This research was supported in part by National Science Foundation grants 0652569 and 0728806, CCF:0430807, and DGE-0504304.

References

1. Adleman, L.M., Kari, J., Kari, L., Reishus, D.: On the decidability of self-assembly of infinite ribbons. In: Proceedings of the 43rd Annual IEEE Symposium on Foundations of Computer Science, pp. 530–537 (2002)
2. Lathrop, J.I., Lutz, J.H., Summers, S.M.: Strict self-assembly of discrete Sierpinski triangles. Theoretical Computer Science 410, 384–405 (2009)
3. Patitz, M.J., Summers, S.M.: Self-assembly of decidable sets. In: Calude, C.S., Costa, J.F., Freund, R., Oswald, M., Rozenberg, G. (eds.) UC 2008. LNCS, vol. 5204, pp. 206–219. Springer, Heidelberg (2008)
4. Rothemund, P.W.K.: Theory and experiments in algorithmic self-assembly. Ph.D. thesis, University of Southern California (December 2001)
5. Rothemund, P.W.K., Papadakis, N., Winfree, E.: Algorithmic self-assembly of dna sierpinski triangles. PLoS Biology 2(12) (2004)
6. Seeman, N.C.: Nucleic-acid junctions and lattices. Journal of Theoretical Biology 99, 237–247 (1982)
7. Wang, H.: Proving theorems by pattern recognition II. The Bell System Technical Journal XL(1), 1–41 (1961)
8. Wang, H.: Dominoes and the AEA case of the decision problem. In: Proceedings of the Symposium on Mathematical Theory of Automata, New York, pp. 23–55. Polytechnic Press of Polytechnic Inst. of Brooklyn, Brooklyn (1962/1963)
9. Winfree, E.: Algorithmic self-assembly of DNA. Ph.D. thesis, California Institute of Technology (June 1998)

Advancing the Deoxyribozyme-Based Logic Gate Design Process

M. Leigh Fanning[1,*], Joanne Macdonald[2], and Darko Stefanovic[1]

[1] Department of Computer Science, University of New Mexico,
Albuquerque, New Mexico, 87131, USA
leigh@unm.edu

[2] Division of Experimental Therapeutics, Department of Medicine,
Columbia University, New York, 10032

Abstract. We previously described a tic-tac-toe playing molecular automaton, MAYA-II, constructed from a molecular array of deoxyribozyme-based logic gates, that uses oligonucleotides as inputs and outputs. We are now developing an ensemble modeling tool for high-throughput oligonucleotide input and logic gate designs. The modeling tool is based on exhaustive reconstruction of both intended and unintended reactions between MAYA-II gates and inputs, and seeks to correlate empirical observations with computational predictions. We present results from computational analysis of the MAYA-II *Yes* logic gate and input interactions. Results indicate that *in silico* modeling correlates with experimental results, creating a predictive value and benchmark. These studies serve purpose towards our goal of constructing a generalized oligonucleotide library for expansion of molecular computation networks beyond hundreds, to millions of potential interactions, conferring greater functionality in terms of both reliability and complexity.

1 Introduction

Computation on a molecular substrate has been physically realized in automata devised by Adelman [1], Benenson et al. [3], and in our laboratory as the MAYA-I [15] and MAYA-II [13] experiments, among others. MAYA, a molecular array of *Yes* and *And* gates, consists of individual deoxyribozyme-based logic gates and sequentially introduced DNA oligonucleotide inputs. These automata effect control by careful biochemical arrangement. With the ensuing heterogenous population of tens of distinct oligonucleotide species together in solution, successful computation depends on achieving satisfactory thresholds of hundreds of desired events, including gates folding to their intended secondary structure and inputs binding to gates at precisely the correct locations along their entire length. These potential events are positive in the sense that they are necessary to carry out the intended computation, yet commensurately there are thousands of negative events that must not occur, or occur at low enough rates so as to minimize noise or system failure.

---

* Corresponding author.

R. Deaton and A. Suyama (Eds.): DNA 15, LNCS 5877, pp. 45–54, 2009.

Design efforts hence walk a tight line. In practice, *in silico* planning [2], [5], [17] has centered around sequence selection that will achieve a particular target secondary structure, match gate recognition regions to the reverse complements of input sequences, and ensure sufficient content difference between all other element combinations that may be present in solution and therefore required to have agnostic relationships. This includes all gate/gate, input/input, and gate/input tuples which could adversely affect reaction product yield and create system noise. Multiple gates and inputs are then combinable, and may carry out concurrent hybridization and cleavage reactions, creating a purely parallel computational system. In setting out to achieve these goals, we assume both negative and positive affinity relationships. By employing Hamming distance metrics, we assume suitable separation will minimize or eliminate the problem of cross-talk. Correspondingly, we assume that matching gate recognition regions to inputs ensures reliable, measurable output products, which are then de facto communication signals carrying information.

A reasonable hypothesis is that these assumptions should be equally borne out in DNA models and further, that empirical results should additionally confirm model predictions. However, as far as we are aware, the empirical feedback from the *in vivo* tuning, testing, and modifications required to achieve computational logic goals via chemical networks has not to date been fully incorporated into the design process. In our experience, the time required to actually make a system work is significant despite use of various design algorithms and predictive models. Too much is left to chance, which translates to an unacceptably long time between system design and test success. We note as well that the field of DNA computing is largely still in demonstration, and that no formal engineering rules have been developed or agreed upon, nor regularly applied in practice.

To these ends, we are undertaking the task of revising the design process. Rather than relying solely on one methodology, or a single analytical model, we will employ a variety of relevant modeling techniques. We will also employ an ongoing mechanism to capture experimental results directly back to the pre-test computational side such that observational knowledge is not lost. To initiate this meaningfully, we are undertaking an analytical reconstruction of the MAYA-II automaton. MAYA-II [13] was designed to play tic-tac-toe against a human opponent. This body of work now serves as a rich source of unmined data which can be further exploited as design feedback. Reconstruction is a reverse engineering effort, and has clear goals. First, we would like to compare and correlate empirical observations to offline model predictions. Second, we want to be able to deduce sources of noise resulting from unintended reactions and explain varying signal intensity from some of the deoxyribozyme gates. We report here development of our general ensemble modeling tool Pyxis, our reconstruction approach, and Phase I reconstruction results which have confirmed some initial expectations. Pyxis exhaustively enumerates all reactions which must be accounted for in a molecular computing system, and allows testing against a suite of existing domain-specific models. Collating model prediction results, along with empirical data comparison, will permit design scaling from the current level of hundreds of competing reactions to millions. In time, we expect a more principled design approach to lead to the development of testable molecular computing benchmarks.

1.1   Development of Pyxis as an Ensemble Modeling Tool

For evaluating and testing different combinations of oligonucleotides which have the possibility of interacting, and evaluating a chemical network from a kinetic perspective, we determined that conceiving of any new DNA domain-specific model or physical simulation was not warranted. Instead, we needed a way to easily account for the prohibitory complexity arising from using different deoxyribozymes and oligonucleotides together in non-trivial ways. Further, we both wished for a more general way to evaluate arbitrary combinations, and observed that greater functionality chemical computing is moving towards more complicated circuits and more circuits networked together. This implies substantially greater numbers of distinct species, which would prove untenable for exhaustive laboratory testing during development. Additionally, inferring function and behavior from nucleic acid structure alone falls short for all but the very simplest of combinations involving only a few participating elements. System parameters are interdependent and must be addressed as closely as possible prior to testing in the laboratory.

Our solution is development of Pyxis, an ensemble modeling tool (Fig. 1) which serves as a framework for enumerating combinations of elements, testing

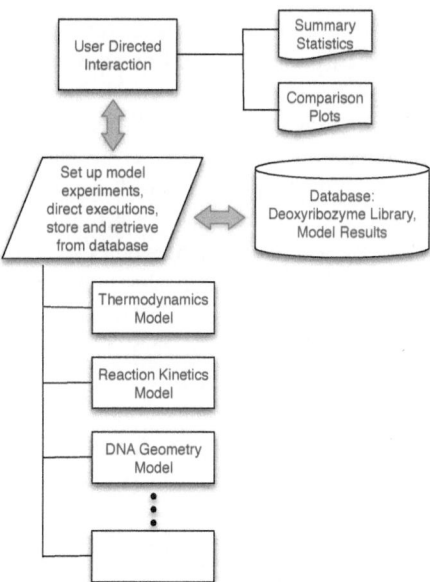

Fig. 1. Pyxis architecture. Single or batch experiments are directed towards requested models. Models remain in natively encoded languages, input and output is packaged by the program.

against a suite of models, collating results and correlating results from different models where possible. Pyxis is encoded as a Python application [12] fronting a PostgreSQL database which houses domain model test results and all deoxyribozyme and oligonucleotide elements. Existing models may be dropped in, input interfacing via command line or files is automatically handled, as is harvesting of data from output files. Users are not required to manage the fine details for each model and are directed to provide only the essential parameters, such as selection of deoxyribozymes from the database, temperature, or similar relevant information. Pyxis brokers all model executions and stores results back into the database for later uses such as plotting, computing statistics, and results comparison where appropriate. The first model incorporated into Pyxis is NUPACK [6]. NUPACK builds on Mfold [17] nucleic acid folding and hybridization prediction software, and extends prediction to an arbitrary number of nucleic acid strands.

1.2 Overview of the MAYA-II Automaton

MAYA-II demonstrated all possible games of Tic-Tac-Toe instantiated in a chemical network. The automaton adversary was able to play each game with a non-losing strategy. Solution wells, representing game squares, were preloaded with deoxyribozyme-based gate constructs designed to carry out the specific Boolean formulae for human play and subsequent automaton response. All gates used one or more stem-loop controllers as the base architecture, each configured using various DNA oligonucleotides. With every input introduction into the solution wells, exactly two gate/input interactions were designed to occur: one in the well where the human intended to mark a square, and the other in a different well where the automaton intended to mark its response. Move detection and display was accomplished via oligonucleotide substrate cleavage by the gate, producing red tetramethylrhodamine (TAMRA) or green fluorescein visible fluorescence. The $\Delta F/min$ signal directly reflected correct hybridization between designated input and gate oligonucleotides. Complete results are in [13]; here we focus instead on the possible interactions between oligonucleotides present in solution. Despite the fact that oligonucleotide content was carefully arranged through string search algorithms, and some thermodynamic modeling was executed prior to testing, getting to reliable, repeatable execution of every game required non-trivial laboratory man-months.

1.3 Handling Combinatorics of Oligonucleotide Populations

A central goal in the reconstruction of MAYA-II, and advancing design methodology, is to account for all potential interactions that could occur incidentally or deliberately within a chemical network. The notion of interaction for stem-loop controllers involves an ordering of certain events. For each event, reconstruction attempts to estimate levels of participation amongst the different nucleic acid reactants, as well as to quantify products. To get from introduction of an input, to detection of fluorescent output, 1) deoxyribozyme gates must correctly fold to the designed secondary structure of stem-loop conformations (Fig. 2), 2) inputs must seek out and bind to the gates in the loop regions connected to stem sections, 3) the stem loop must undergo a conformational change wherein it opens, and 4) the exposed remainder of the stem section must bind and cleave a substrate oligonucleotide. The critical steps are gates attaining expected secondary structure and inputs correctly binding to gates, each of which previously was designed on the basis of minimization of predicted free energies of the gates and gate/input complex formations. Other relevant factors, which are no less important, are gate and input concentrations, diffusion, temperature, k_{on} and k_{off} binding and unbinding efficiencies, and the chemical kinetics involved for either hybridization or cleavage reactions. These are beyond the scope of this paper, and will be addressed in the ongoing evolution of our ensemble model.

We have initially focused our efforts on accounting for all possible ways gates and inputs might have come together, as either positive in terms of one or more inputs correctly binding to a gate which has folded properly, or negative for any other combination, including inputs binding to other inputs, gates binding to other gates, or inputs binding to the wrong gates. For gate/input combinations we consider the set of all games, all plays within each game, and all wells since the same input was introduced into each well at each play:

$$wells = \{1,2,3,4,6,7,8,9\}, games = \{1,\dots,76\}, plays = \{1,2,3,4\}$$

$$I = \{input_{i,j,k} | i \in wells, j \in games, k \in plays\}$$

$$G_{\text{YES}} = \{gate_{i,j,k_1} | i \in wells, j \in games, k_1 \in plays\}$$
$$G_{\text{AND}} = \{gate_{i,j,k_1,k_2} | i \in wells, j \in games, k_1, k_2 \in plays\}$$
$$G_{\text{ANDANDNOT}} = \{gate_{i,j,k_1,k_2,k_3} | i \in wells, j \in games, k_1, k_2, k_3 \in plays\}$$

$$G = G_{\text{YES}} \cup G_{\text{AND}} \cup G_{\text{ANDANDNOT}}$$

With this accounting we make explicit the gate constructions where *Yes, And,* and *AndAndNot* gates respectively admit one, two, and three inputs. We evaluate reaction sets I, the set of inputs alone, G, the set of gates alone, $G \times G$, the set of pairs of gates, $G_{\text{YES}} \times I$, the set of pairs of *Yes* gates and inputs, $G_{\text{AND}} \times I \times I$, the Cartesian product set of *And* gates and input pairs and $G_{\text{ANDANDNOT}} \times I \times I \times I$, the Cartesian product set of *AndAndNot* gates and input triples. The strategy treats inputs as multisets, and therefore considers the same inputs binding to multiple sites with the two and three input gates. We are not composing and testing beyond these limits unless particular oligonucleotides reveal themselves to be highly reactive, which disallows full cross product sets of all participants. Combinations are deduced by Pyxis, rather than by hand, and are generalized such that any solution of oligonucleotides may have potential interactions enumerated.

2   Results

For Phase I reconstruction, we determined results for MAYA-II *Yes* gates using Pyxis to form evaluation sets and test each using NUPACK. Two different deoxyribozyme designs were used in the development of the *Yes* gates. Human and MAYA moves respectively employed the 10 nucleotide (nt) 8.17.1 and 6-8 nt E6 stem-loop structures (Fig. 2). Each presents an input recognition region in the form of the 15 nt loop segment. The stem region, alternatively termed the critical structure, must pair-bind as shown such that no gaps, puckers, or otherwise ill-formed bindings occur, leaving the loop exposed. The loop region must itself be free of extensive pair bindings. Attaining these configurations constitutes optimal structure folding. All the *Yes* gates were analyzed individually for self-hybridization and optimal folding. The evaluation shows all gates folding properly within the critical structure, over half (24/40 = 60%, Tab. 1) show small amounts of additional binding within the loop region intended for input recognition. These additional bindings use between one and five pairs of nucleotides within the loop region, and were observed for both stem-loop types. Hybridization evaluations of gates with intended inputs constituted a positive set, and cross-hybridization evaluations of gates with unintended inputs constituted a negative set. For MAYA-II, 40 *Yes* gate combinations were positive and 1240 were negative. The predicted minimum free energy (MFE) structures (Tab. 2) show distinct differences in average MFE between the gates alone and gate/input combinations. This serves as a predictive cut-off value for indicating gate activation, whether through intended or unintended input binding. Some

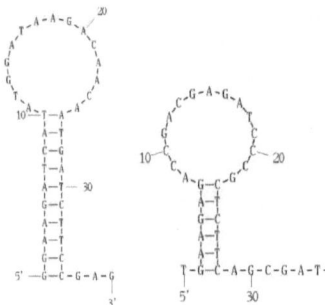

Fig. 2. Stem-loop portions (DinaMelt server, [14]) of Yes-Gate deoxyribozymes. The 8.17.1 design (left) uses a 10 nt stem whereas the E6 design (right) nominally uses a 6 nt stem. In some cases input dependent adjustments of the E6 resulted in 7 and 8 nt stems. Not shown past nucleotide 30 is the core region used for secondary substrate cleavage reactions.

Table 1. Number of nucleotide pairs bound within gate input recognition regions

status	clean	1 pair	2 pair	3 pair	4 pair	5 pair
count	16	2	14	5	2	1

Table 2. Minimum Free Energy Secondary Structures

(kcal/mol)	8.17.1 alone	8.17.1 positive	8.17.1 negative	E6 alone	E6 positive	E6 negative
Average	-12.17	-28.30	-19.06	-12.45	-31.48	-19.71
St. Dev.	0.65	1.76	1.45	0.81	2.12	1.96
Minimum	-13.74	-32.76	-23.63	-14.54	-34.83	-26.30
Maximum	-11.21	-24.23	-16.19	-11.88	-28.23	-15.73

cross-talk is suggested with the high end of positive sets approaching the low end of negative sets. MFE analysis of concurrent input and gate binding had not previously been performed during MAYA-II engineering. Additional data is required to reconstruct likely product populations as the most stable structures, as reported via MFE evaluation, does not indicate time of formation or possible domination by less probable but faster binding structures. Cleanliness of gate-input bindings for design combinations and verification of lack of affinity for incidental, unintended gate-input bindings is observed for 8.17.1 stem-loop designs generally, and only partially for E6 stem-loop designs. Space limitations prevent full illustration in this paper. The data however supports empirical laboratory observations that the shorter stems of the E6 designs were more prone to incorrectly formed gate-input complexes and exhibited far more problems with cross-talk bindings between gates and unintended inputs.

3 Discussion

During the MAYA-II development time, the design process only allowed for limited assessment of the system as a whole. Gates were manually designed by modeling an individual deoxyribozyme structure with 15 nt sequences of interest inserted into one or more input binding regions depending on gate type. Free energy of one or more input bindings had never before been determined, nor was the system evaluated in terms of all potential reactions other than bioinformatics-based filtering to ensure sequence separation. During the experimentation process, time-consuming effort was expended on adjusting gate concentrations, and making small incremental changes to the basic gate designs in order to fine-tune the fluorescent signal output and avoid noisy combinations. These changes were based on laboratory experimentation alone and involved replacing inputs in a few cases, reversing loops, and adding or removing single nucleotides. Part of our reconstruction involves evaluating each of these incremental changes analytically, and then comparing directly with previously gathered empirical observations. This will allow us to see how well model predictions correlate to incremental laboratory engineering results, or, as the case may be, where the correlation fails. In both respects we obtain feedback not only on model predictiveness, but also on the effectiveness of minute modifications in analytical terms.

These critically important, but essentially very fine details, are now becoming incorporated into a high-throughput approach to deoxyribozyme gate design. Our process is general such that millions of sequences can be tested. We are already employing this expanded design methodology to study optimal binding sequences in the creation of deoxyribozyme-based virus sensors, including sensors for West Nile Virus and other Select Agent List viruses [18]. Moreover, our evaluation framework is modular, and designed to accommodate additional non-thermodynamic models which will augment understanding of the underlying physical processes at work. Two additional fully functional models have been identified for near-term inclusion: Kinefold [16] and Copasi [11]. The former will be directly comparable with NUPACK secondary structure prediction results, and allows for pseudoknots which is forbidden by NUPACK. The latter is intended to bridge the missing gap of reaction kinetics modeling. We are actively working to tie new modeling results with empirical evidence. To the best of our knowledge, this approach has not been undertaken elsewhere and represents a verifiable, concrete path towards large-scale whole-system design. Pyxis is a dynamic tool, which can and should grow as different designs are proposed and evaluated.

We have answered here several basic questions for the simplest of MAYA-II gates. *Yes* gates fold properly while exhibiting some variable binding tendencies within open loop regions. The input binding regions were designed to provide accessible loops. Hence, we expect collection of loop secondary structure information to be useful in future gate designs, since intra-loop binding prevents full access to inputs. The assumption that gate recognition regions and inputs bind completely for intended inputs, and show no binding or activity for unintended inputs, is only partly borne out by these results. Partial gate/wrong input binding, a negative combination, could be allowing some percentage of in-solution gates to be incorrectly occupied and creating enough force such that some degree of loop opening is occuring and exposing part of the stem to the substrate strands employed in the secondary fluorescent signal reactions. Model results show variability in binding with negative inputs, which corresponds to experimental data generated during the original MAYA-II construction in two ways. First, E6 gates required more manipulation of the stem region, as well as higher concentrations to obtain clear digital behavior. Second, occasionally, spurious signaling was noted early on after input introduction. These false signals tended to die down, and were later overtaken by the intended signals. Pyxis generated results confirm that use of the E6 stems may have been partly responsible for some of this behavior. The extra three nucleotides in the 8.17.1 stem yielded reliable results for both positive and negative combinations.

Our results additionally question the assumption that complete binding along the entire 15 nt recognition length is required for gate activation. Quantification of this length, and mapping out which positions along the region is sufficient, is currently being determined experimentally. These results will be captured back into Pyxis. New project needs call for possibly increasing the loop sizes, or changing how recognition can be done with inputs longer than 15 nt. In the longer term, this work serves purpose in developing gates which can be pulled off the shelf and readily combined to suit logic goals; in essence it enables the development of a reusable oligonucleotide library.

3.1   Assessing Modeling Approaches

Historically, the basic design problem for molecular computing with nucleic acids has centered around sequence selection to achieve a particular target secondary structure. Design methodologies dictate maximal probability of target structure stability based on energy considerations. Considering minimum free energy predicted structures or complexes is the necessary first step for thermodynamic evaluation of arbitrary oligonucleotide interactions. The next step we are taking, however, is consideration of the remaining probability mass of possible complex formations which exhibit slightly higher binding energies. The MFE complex could in reality be dominated by the remaining possible complex formations, any one of which might not lead to the desired reaction, either in the positive or negative sense. We note that theoretical models such as DNA code words and process algebras, [2], [4], [9], which strive to formally codify notions of biological function strictly based on Watson-Crick base-pairing, do not capture the underlying jostling between competing structures which must be accounted for in the laboratory to actually make things work.

Outside of thermodynamics, there are other modeling options and optimization possibilities, some or all of which can be codified into design rules. Strand secondary structure specific considerations (see [5]) are GC content, Shannon entropy, and symmetry minimization. GC and AT bonds are treated as equals, yet physically they are not. Strand tertiary structure considerations include DNA curvature, and base pairing and stacking where strand geometries [8], are taken into account. Finally, the chemical kinetics of folding, cleavage, and hybridization [7], dictate how fast or slow strand interactions occur. Ultimately, our goal in development of an ensemble model, is to infer a weighting scheme for all factors determined to be the most relevant over the whole of a system. This is a departure from old methods that rely on energy considerations alone and consider a chemical network as a sum of individual non-interdependent physical effects. Exploitation of this understanding suggests alternative control mechanisms as well where the energy landscape is intentionally kept highly variable, binding affinities are calibrated with deliberately inserted mismatches, or faster reactions are introduced to control the amount of products available for later reactions.

Acknowledgements. The material is based upon work supported by the National Science Foundation under NSF Grants 0829881 and 0829793. MLF is grateful for support from Elsa Garcia and the UNM Intel Minority Engineering Scholarship.

References

1. Adleman, L.: Molecular computation of solutions to combinatorial problems. Sci. 266, 1021–1024 (1994)
2. Andronescu, M., Dees, D., Slaybaugh, L., Zhao, Y., Condon, A., Cohen, B., Skiena, S.: Algorithms for testing that sets of DNA words concatenate without secondary structure. In: Hagiya, M., Ohuchi, A. (eds.) DNA 2002. LNCS, vol. 2568, pp. 182–195. Springer, Heidelberg (2003)

3. Benenson, Y., Gil, B., Ben-Dor, U., Adar, R., Shapiro, E.: An autonomous molecular computer for logical control of gene expression. Nat. 429, 423–429 (2004)
4. Cardelli, L.: Strand algebras for DNA computing (2009),
   http://lucacardelli.name/Papers
5. Dirks, R., Lin, M., Winfree, E., Pierce, N.: Paradigms for Computational Nucleic Acid Design. Nucl. Acids Res. 32(4), 1392–1403 (2004)
6. Dirks, R., Bois, J., Schaeffer, J., Winfree, E., Pierce, N.: Thermodynamic Analysis of Interacting Nucleic Acid Strands. SIAM 49(1), 65–88 (2007)
7. Flamm, C., Hofacker, I.: Beyond energy minimization: approaches to the kinetic folding of RNA. Chem. Mon. 139(4), 447–457 (2008)
8. Harvey, S., Wang, C., Teletchea, S., Lavery, R.: Motifs in Nucleic Acids: Molecular Mechanics Restraints for Base Pairing and Base Stacking. Jour. Comp. Chem. 24(1), 1–9 (2003)
9. Heitsch, C., Condon, A., Hoos, H.: From RNA Secondary Structure to Coding Theory: A Combinatorial Approach. In: Hagiya, M., Ohuchi, A. (eds.) DNA 2002. LNCS, vol. 2568, pp. 215–228. Springer, Heidelberg (2003)
10. Hinsen, K.: The Molecular Modeling Toolkit: A New Approach to Molecular Simulations. Jour. Comp. Chem. 21(2), 79–85 (2000)
11. Hoops, S., Sahle, S., Gauges, R., Lee, C., Pahle, J., Simus, N., Singhal, M., Xu, L., Mendes, P., Kummer, U.: COPASI - A Complex Pathway SImulator. Bionfrmtcs. 22, 3067–3074 (2006)
12. Langtangen, H.: Python Scripting for Computational Science. Springer, Heidelberg (2004)
13. Macdonald, J., Yang, L., Sutovic, M., Lederman, H., Pendri, K., Lu, W., Andrews, B., Stefanovic, D., Stojanovic, M.: Medium Scale Integration of Molecular Logic Gates in an Automaton. Nano. Ltrs. 6(11), 2598–2603 (2006)
14. Markham, N., Zuker, M.: DINAMelt web server for nucleic acid melting prediction. Nucl. Acids Res. 33, W577–W581 (2005)
15. Stojanovic, M., Stefanovic, D.: A deoxyribozyme-based molecular automaton. Nat. Biotech. 21(9), 1069–1074
16. Xayaphoummine, T., Bucher, T., Isambert, H.: Kinefold web server for RNA/DNA folding path and structure prediction including pseudoknots and knots. Nucl. Acids Res. 33, 605–610 (2005)
17. Zuker, M.: Mfold Web Server for Nucleic Acid Folding and Hybridization Prediction. Nucl. Acids Res. 31(13), 3406–3415
18. National Select Agent Registry, http://www.selectagents.gov

DNA Chips for Species Identification and Biological Phylogenies

Max H. Garzon, Tit-Yee Wong, and Vinhthuy Phan

Biology, Computer Science, The University of Memphis, Tennessee 38152
{mgarzon,tywong,vphan}@memphis.edu

Abstract. The codeword design problem is an important problem in DNA computing and its applications. Several theoretical analyses as well as practical solutions for short oligonucleotides (up to 20-mers) have been generated recently. These solutions have, in turn, suggested new applications to DNA-based indexing and natural language processing, in addition to the obvious applications to the problems of reliability and scalability that generated them. Here we continue the exploration of this type of DNA-based indexing for biological applications and show that DNA noncrosshybridizing (nxh) sets can be successfully applied to infer *ab initio* phylogenetic trees by providing a way to measure distances among different genomes indexed by sets of short oligonucleotides selected so as to minimize crosshybridization. These phylogenies are solidly established and well accepted in biology. The new technique is much more effective in terms of signal-to-noise ratio, cost and time than current methods. Second, it can be scaled up to newly available universal DNA chips readily available both *in vitro* and *in silico*. In particular, we show how a recently obtained such set of nxh 16-mers can be used as a universal coordinate system in DNA spaces to characterize very large groups (families, genera, and even phylla) of organisms on a uniform reference system, a veritable and comprehensive "Atlas of Life", as it is or as it could be on earth.

Keywords: DNA codeword design, noncrosshybridizing oligonucleotide bases, phylogenetic analysis, genomic signatures, DNA chips, 16S rRNA tree of life.

1 Introduction

The field of DNA computing has originated novel ideas and uses for DNA as a computational medium (Adleman, 1994) and as smart glue for a variety of self-assembly applications (Seeman 2003; Winfree et al., 1998). These applications have stimulated the development and analysis of various techniques aimed at finding large sets of short oligonucleotides with noncrosshybridizing (nxh) properties by several groups, particularly the PCR Selection (PCRS) protocol used below (Garzon et al., 2009; Deaton et al., 2006; Tulpan et al., 2005; Chen et al., 2006). In turn, these sets have found new applications, such as novel DNA approaches to natural language processing (Neel at al, 2006; Bobba et al., 2006) and DNA-based memories (Neel & Garzon, 2008). In this paper, we continue the exploration of these new tools and show how recent solutions to the codeword design problem in DNA Computing (Garzon et al.,

R. Deaton and A. Suyama (Eds.): DNA 15, LNCS 5877, pp. 55–66, 2009.

2009; Bobba et al., 2006; Deaton et al., 2006) can be used as the foundation of new methods to establish biological phylogenies based purely on genomic DNA. After a review of pylogenetic methods in Section 2, the foundations of the method developed in prior work is summarized in Section 3.1. Section 3.2 describes the new method for phylogenies from genome wide data alone, including its validation in reproducing *ab initio* a well established and accepted phylogeny in biology, the 16S rRNA tree of life, from genomic data alone. In Section 4, we discuss some of the significance of these results and some implications for future research.

2 Species Identification

In this section we summarize the basic ideas about phylogenetics. We focus on the aspects relevant to the novel methods to be described later in Section 3.

Phylogenetics is the study of evolutionary relatedness among various groups of organisms such as species, genera, families, orders, classes, phyla, kingdoms and domains (Henning, 1950). It is one of the forms of taxonomy, the classification of organisms according to general similarity. The traditional methods of taxonomy and phylogeny are based on similarity measurements of many observable features. Modern phylogenetics often uses a certain profile of the DNA sequences on the chromosomes to provide evidence of the relatedness of living organisms. Using DNA, the blue print of life, as a primary record of evolution has proved to be far superior than any traditional methods of taxonomy. The most commonly used methods to infer DNA relatedness between species include parsimony, maximum likelihood, and MCMC-based Bayesian inference. The general technique is to use some metric to estimate the similarity between organisms and construct trees based on approximate phylogenetic relationships, usually based on molecular phylogeny where the characters are aligned nucleotide or aminoacid sequences. Evolution is regarded as a branching process, whereby individuals may undergo genetic changes that may diverge into separate species. Some of these new species might successfully reproduce and continue to diverge into newer species, while others might go extinct. These genetic changes over millions and millions years were not only influenced by the rate of DNA changes but also reflected the selection forces imposed by natural environments. Thus evolution may be visualized as a multidimensional feature-space that a population (species or genus, for example) moves through over time.

There are two major approaches to examine phylogenetic relationships. So-called **Phenetic** methods are based on more or less arbitrary characteristics available to the analyst, while **cladistic** methods are based on the assumption that members of a group share a common evolutionary history. Organisms are more closely related to one another if they share a set of unique features that were not present in distant ancestors but which are shared by most or all of the organisms within the group. Most of today's evolutionary biologists favor cladistics, although a strictly cladistic approach may result in debatable and even counterintuitive results. (NCBI, 2008 http://www. ncbi.nlm.nih.gov/About/primer/phylo.html).

An organism can inherit genes in two ways: from a parent (so-called vertical gene transfer), or by horizontal or lateral gene transfer, in which genes "jump" between unrelated organisms, a common phenomenon in prokaryotes and many viruses. In order to come up with robust models, biologists base their phylogenies on so-called most conserved genes that remain unchanged in vertical and lateral transfer. Conserved genes often served a vital function of the cell. Slight changes in the sequences of these genes would often result in cell death. Of these, ribosomal RNA (rRNA) genes are the most commonly recognized; these genes are needed for the formation of the ribosome—a large protein complex composed of more than 50 proteins and peptides. Ribosomes primarily serve to synthesize proteins for the cells. In addition, after organizing the ribosomal protein complex, rRNAs in the ribosomes are still needed to direct the flow of protein synthesis, checking the proper alignments of the two subunits of the ribosomes, the mRNA and the tRNA to insure the production of the right proteins. Because of the multiple functions of rRNA genes in the formation of ribosomes and in protein synthesis, beneficial mutations in the rRNA genes were rare. Since all living organisms have to make protein, and ribosomes were ubiquitous, rRNA genes are therefore commonly recommended as molecular clocks for reconstructing phylogenies. This approach is particularly useful for the phylogeny of microorganisms such as bacteria, which are morphologically too simple to be classified based on phenotypic signal, i.e. visible traits (characters, or features). Among the three types of rRNA genes, the 16S rRNA gene set is the most commonly used. (http://en.wikipedia.org/wiki/Phylogenetics, 2008).

2.1 Identifying Single Individuals: Biomarkers

Every response of a living cell to its environment can ultimately be traced to the genomic information encoded in DNA, albeit by a complex process critically regulated by additional factors and conditions. It is also well known that great segments of DNA are "junk", i.e., not directly related to a particular function. Many of them were likely remains of past, dead-end evolutions, others were pseudogenes that gradually deteriorated and lost their original functions as the environment changed. The so-called central dogma in modern biology and the success of multiple genome projects pose the challenge to identify and map those DNA regions coding for differences evident among biological organisms, in other words, to construct a "Rosetta stone" of biological phylogeny.

This problem has been addressed by biologists in trying to identify critical DNA segments for species identification, for example. The most obvious and articulate program to-date to provide the taxonomic community this type of tools and techniques has been the judicious selection of molecular markers for species-level identification, or so-called *barcodes*. DNA barcoding is a technique for characterizing species of organisms using a short DNA sequence from a standard and agreed-upon position in the genome. DNA barcode sequences are very short relative to the entire genome and they can be obtained reasonably quickly and cheaply. For example, "the cytochrome c oxidase subunit 1 mitochondrial region (COI)" is emerging as the standard barcode region for higher animals. It is 648 nucleotide base pairs long in most groups, a very short sequence relative to 3 billion base pairs" of nucleotides present in the human genome, for example. (http://www.barcoding.si.edu/DNABarCoding.htm). Identification is performed by spotting or synthesizing, on a microarray (Stekel, 2003), the Watson-Crick complements of the distinguisher strands, and then hybridizing to the microarray the fluorescently labeled DNA extracted from the unknown microorganism. The Barcode Initiative aims to construct a public reference library of species identifiers which could be used to assign unknown specimens to known species. Nearly optimal theoretical and practical algorithms have been developed to barcode given genome families (Zhou et al., 2008; Dasgupta et al., 2005).

As useful as it might appear for phylogenetic purposes, a biomarker is far from ideal. What is desirable is a more comprehensive set of features describing the biological functions that characterize individuals in a given set of organisms. More recent and systematic molecular approaches have been developed, but, although they provide distinctive biomarkers, they are not often ubiquitous in all genes. These methods are, nevertheless, generally congruent with the results of the 16S rRNA based phylogeny (discussed below in Section 2.2.) The rapid accumulation of complete sequences of a number of prokaryotic genomes has made it possible to analyze the relationships between organisms at the whole-genome level in some groups. New biosignatures have provided independent assessment in taxonomies and, in some cases, have improved the accuracy of the single-gene based 16S-rRNA system (Zhou et al., 2008; Henz et al., 2005). However, most of the approaches are still relatively subjective and far from being a universal barcode representing *all genes in all genomes*.

The new technique presented in Section 3.2 can be regarded as a generalization of this concept of biomarker so as to include not only uniquely DNA identifiers, but as to capture a more comprehensive picture of the biological relationships at play in a given family or genus, not only in terms of a genome-wide method, but, more importantly, as a method potentially scalable to all organisms. To evaluate the new method in Section 3, some preliminaries are required and presented next.

2.2 Phylogeny by Conserved Features and the 16S rRNA Tree of Life

The most general approach of designating taxonomic ranks is the clustering by 16S rRNA gene data. However, this method is not ideal. No standards exist for measuring the relatedness distances between genomic clusters; furthermore, others factors could prevent clear-cut separation between clusters, for example, multiple copies of 16S-rRNA genes in a single isolate, the possibility exists of 16s-rRNA being horizontally transferred from one species to the others, and differences in mutation rates in different species due to different environments and mutation saturation. The objectivity of selecting conserved proteins as a guiding tree for constructing the otherwise un-rooted 16S-rRNA tree (Woese, 2002), and the different mathematical models employed in clustering, all contribute to the inconsistency in taxonomic rankings. Furthermore, although all living organisms have the 16S rRNA genes, these genes might not reflect the natural history of a majority of the genes in the species' genome and thus violate the "ubiquitous rule" within the genome of the species (Margulis, 1993). Additionally, many prokaryotes are known to carry multiple chromosomes and plasmids (e.g. *Azotobacter vinelandii, Deinococcccus radiodurans*). The origins of these chromosomes and plasmids remain elusive and cannot be resolved by the single-gene approach. Consequently, grouping of clusters are more often based on peculiar intuitions of individual researcher(s) in specific families or genera. For the evaluation of our method, we use an expansion of this generally accepted scheme, the 16S rRNA tree (a fragment is shown Fig. 3, left), in order to compare organisms outside the

Table 1. A selection of six (6) pervasive and important bacterial organisms selected for evaluation of the new pylogenetic analysis described in Section 3. Their nucleic acid files were downloaded in FASTA form and a phylogenetic analysis made to produced a tree, using JavaTreeView, based on a refinement of the 16S rRNA tree of life, originally constructed on the analysis of highly conserved sequences by traditional methods in phylogenetic analysis.

Organism	Genome Size (Mbytes) / ORFs selected (Kbytes)
Escherichia coli CFT073	5.05Mb / 510K
Escherichia coli K12	4.37Mb / 438K
Photobacterium Profundum	3.59Mb / 186K
Pseudomonas Aeruginosa PA01	5.89Mb / 192K
Salmonella Enterica Choleraes.	4.51Mb / 151K
Vibrio. Fisheri ES114	2.64Mb / 179K

region of highly conserved genes. The expansion builds on the suggestion that evolution and speciation were facilitated by genome expansion. Further details on this method and the construction of this tree can be found in (Wong et al., 2008).

To illustrate how we the methods works and performs in practice, the FASTA Nucleic Acid files of six bacterial genomes were downloaded from The Technology Institute for Genomic Research (http://cmr.tigr.org) as shown in Table 1. The occurrences of stop sequences in the first, second and third reading frames of each open reading frame (ORF) from a chromosome were counted by a program written in C. The resulting counts of each in each reading frame, Cistronic Stop Signals (CSS), together with the gene IDs of the corresponding organism were transferred to an Excel spreadsheet and were converted to their corresponding partition functions – termed CSRS, using the built-in formulas in Excel. The hierarchical clustering algorithm in Cluster 3.0 was used for correlation clustering of the CSRSs from different chromosomes. Phylogenetic trees were constructed by JavaTreeView. Both Cluster 3.0 and JavaTreeView were downloaded from Michael Eisen's site (http://rana.lbl.gov/EisenSoftware.htm). The results are shown in Table 1 and the corresponding tree 16S rRNA in Fig. 3 (left).

3 DNA Indexing on Universal DNA Chips

In this section we describe the new method for taxonomical classification based on genomic-wide DNA alone. The foundations of the method have been established in prior work (Garzon and Phan, 2009; Garzon et al., 2004b; Bi et al., 2003), but the main features are summarized here to make this paper self-contained.

Currently, the most powerful way to capture genomic data is on DNA microarrays (Stekel, 2003) and DNA chips. The fundamental problem with standard DNA microarrays is that useful *information* is often deeply buried under mountains of "noisy" data. In addition, the questions on the confidence, accuracy and reliability (false positives/negatives) of the analyses become even more prominent in applications such as species identification and diagnosis.

On the other hand, a thread of work in the field of DNA computing (Garzon and Yan, 2008) has carried out deep analyses of oligomer spaces in solving the so-called codeword design problem (Garzon et al, 2009; Qiu et al., 2008; Garzon et al., 2006;

Tulpan et al., 2005). The original issue was the tendency of DNA oligos to hybridize in undesirable ways when used for solving computational (Adleman, 1994), for self-assembly (Reif et al., 2001), for DNA memories (Neel & Garzon, 2008), and also for DNA microarrays (Bobba et al., 2006) and text processing (Neel & Garzon, 2006). Rigorous analyses have been conducted for a theoretical framework, their sensitivity and reliability, as well as their information capacity (Garzon et al., 2009; 2006; 2005; Bi et al., 2003). Nearly optimal designs have been provided, sometimes through tours de force requiring nearly years of supercomputing time, as is the case of the 16-mer prototypes used below in the universal DNA chip (Garzon et al., 2009; 2006). We summarize these results next.

3.1 Indexing Using a DNA Basis

In this section, we summarize the theoretical foundations of the metric used in our phylogenetic analysis. If DNA chips are to produce highly informative "pixels" (spots), a principled study of what we term "DNA spaces" would help. We conducted such a study in prior work (Garzon et al. 2006; 2005, 2004b). DNA chips usually require uniform length short oligos of up to $n=100$-mers or so. The magic behind the operation of a DNA chip is hybridization based on strand homology (i.e. Watson-Crick complementarity.) Let B be a set of DNA n-mers affixed to the chip. These oligonucletoides will be called *probes* (or in the case of a very special type of set, an encoding basis, to be defined precisely below.) In practice, this set is a judicious se-lection of (the complements) of some fragments of some target gene(s) from a target organism, or even a full selection of the target organism (as in the case of a DNA microarray.) A given (possibly unknown) target is digested, usually tagged, and poured over the chip. A signal is produced by a pattern of hybridization to the probes on the chip. If the oligo probes on the chip are not pre-processed, the result will be a signal that can be highly variable and essentially unreproducible upon repetition of the readout. The corresponding payoff will thus be noisy and uncertain and the result-ing analyses must thus be expected to be difficult at best, unreliable at worst. To make the point with an extreme case, B could consist of hairpins that will hybridize to themselves and produce no signal whatsoever, when it fact there may exist useful information in the given targets; in a more common case, they could be duplicates or small mutations of one another (perfect match/single mismatch), presenting a gradient pattern capturing a single bit of information over several spots. Is it really possible to eliminate or reduce the noise in the signal, while at the same time increase both the reliability and the sensitivity of the chip? How?

(Garzon et al 2005) have demonstrated that these goals can be achieved by pre-processing of the oligo probes. The key property is named *noncrosshybridization* (nxh), which comes in degrees of quality determined by a numeric parameter ρ that controls for hybridization stringency of reaction conditions. The best choice would be the Gibbs energy, or approximations thereof (Watkins & SantaLucia, 2005), typically set at -6 Kcal/mole (the minimum free energy required for hybridization of two strands to occur.) The concept is best understood by an analogy with more familiar concepts in ordinary Euclidean spaces. If, by way of analogy only, the full space of m-mers in DNA space were to be displayed as a region in a Euclidean place so that the distance between them was proportional to their degree of Watson-Crick complementarity, a noncrosshybrizing set would be an "orthogonal" set of points that are "hybridization independent", i.e., have no redundancies in terms of hybridization. On a noncrosshybridizing chip, a random target digested to fragments of comparable probe size is much more likely to hybridize to fewer probes, and under appropriate stringency ρ, *to at most one probe*. This so-called *nxh property* immediately translates into the desirable properties to the problems mentioned above. The noise is notably reduced (in fact, it is completely eliminated under ideal conditions), results will be more predictable, and the corresponding analyses will be much more reliable, as explained next.

The second key property of the DNA chip is to be *complete*, i.e., so that an arbitrary random n-mer will hybridize to at least one (and hence exactly) one probe, for a given level of hybridization stringency ρ. For example, under most stringent conditions (highest nxh quality), any target will only hybridize to its perfectly complementary strand, so a complete set will be about half the size of the DNA space (we obviously need to exclude palindromic probes.) This size is prohibitively large and unrealistic in terms of reaction conditions. At the opposite extreme (lowest nxh quality), any target hybridizes to many (or even all) probes, so that the nxh basis code consists of only one probe, a situation equally undesirable. The appropriate set is somewhere between these two extremes. These good nxh bases for a given set of 16 and 20-mer probes have been extracted from the vastness of genomic or DNA spaces after searches that have extended for long ;periods of time (Garzon et al., 2009; Garzon et al., 2006).

The *genomic signature* of a target genome X with respect to chip B is a vector $x=\{x_i\}$ of size m (the size of the chip, or number of distinct oligo probes in B), obtained by shredding X to fragments of length comparable to n, the size of the oligos in B, and pouring them over B under appropriate reaction conditions. The intensity value x_i of each pixel i is proportional to the number of labeled fragments that hybridize to probe i. The vector x can then be visualized as a 2D picture, as illustrated in Fig. 1 below. For example, the chip B is noncrosshybriding (under stringency ρ) if the signature of each and every one of the complementary probes poured as targets strands has a point signal of maximum intensity, i.e., no two different strands in B will hybridize to one another under reaction conditions ρ. It will be important for the experimental results below to note that, as is easily seen from the definition and an nxh basis, ideal genomic signatures on nxh chips obey a superposition principle under ideal conditions, i.e., the signatures of the superposition of two genomes is identical to the probewise superposition of the ideal genomic signatures of the two component genomes. This superposition in only approximate under more relaxed stringency conditions. Further details about nxh chips can be found in (Garzon et al., 2009; Bobba et al., 2006; Garzon et al., 2004b).

Further theoretical work has shown that a non-crosshybridizing (nxh) chip of single-stranded probes (also referred to as DNA codesets, or DNA chips), has a number of distinct and powerful advantages in terms of the signal they generate (Garzon et al., 2005, 2004b). First, it will have a superior level of sensitivity and resolution for the same domain compared to standard microarrays, simply because minimizing cross-hybridization between basis probes reduces the number of probes a target can hybridize to (at most one in the optimal scenario.). Therefore, signatures will exhibit minimum variability and noise, assuming saturation conditions in target concentrations and long enough relaxation time to obtain a full signal (order of hours). This result has been already demonstrated in simulation in (Bobba et al., 2006) with a selection of genes important in seven human diseases. We emphasize once again that this reduction in the number of probes is typically considered a loss of redundancy, and hence a loss of signal-to-noise ratio in standard analyses (Stekel, 2003), that would diminish the payoff of the readout. The results below provide additional evidence to the counterintuitive fact that, on the contrary, this judicious selection of probes will improve the sensitivity of a chip for target genome discrimination, this time in the field of phylogenetic analysis.

3.2 Building Phylogenetic Trees *ab initio* Using DNA Indexing

In this section, we describe the new metric to build phylogenetic trees mentioned above. Once the genomic signatures are obtained, distances between them could be computed using standard distance measures, such as Euclidean metrics, cosine, or correlation, as shown in previous analyses on DNA microarrays (Bobba et al., 2006). The validation of these trees has been conducted in evolutionary biology, by cross-validating *ab initio* taxonomies obtained by comparative analyses of genomic signatures with well established and accepted taxonomies such as the Bergey's 16S rRNA relatedness of these organisms described above.

The genomic signatures described below have been obtained using a simulation tool, a virtual test tube, *Edna*, developed by the first author in last few years for simulating the DNA reactions such as hybridization, ligation, self-assembly, and enzymatic reactions (Garzon et al., 2004; Blain et al, 2004) *Edna* has produced results highly correlated, if not identical, to the results of experiments *in vitro*, including a successful run of Adleman's solution to the difficult Hamiltonian Path Problem (Adleman, 1994) in silico for graphs up to 12 vertices producing error-free solutions (Garzon et al., 2003). Therefore, there is good evidence that Edna will produce high reliable estimates of the DNA reactions and other events in a wet tube.

Using as a seed pool the same selection of ORFs from these organisms used to obtain the extended 16S rRNA tree (Table 1), we ran the PCR Selection protocol in simulation to obtain a chip of about 1600 16-mers for these targets by the methodology described in (Garzon et al., 2009; Deaton et al., 2006; Chen et al., 2006). Fig. 1 shows the resulting genomic signatures of the bacteria given in Table 1.

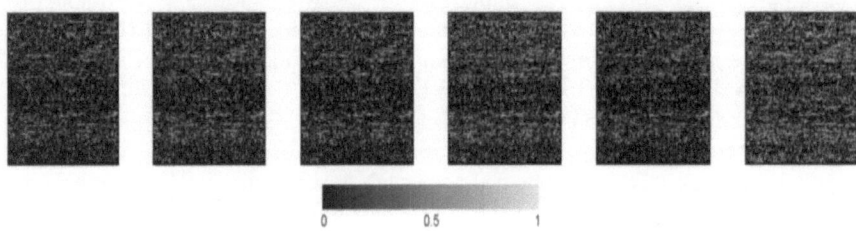

Fig. 1. Genomic signatures of six bacteria (permuted from Table 1) on a nxh DNA chip obtained by PCR Selection from a seed pool of shredded ORFs of their chromosomes. The signatures show clearly significant relative differences and can be contrasted by objective measures to build a phylogenetic tree. Furthermore, probes with most significant similarities and differences can be extracted by pixelwise comparison of pairs of genomic signatures.

The signatures show a clear difference between the target genomes, not only visually to the human eye, but through more objective measures. The obvious measure, the
plain Euclidean distance between signatures vectors in Euclidean space, already shows
a high degree of correlation comparable to the 16S rRNA tree. Nonetheless, we used a
more refined measure in our analysis and produced a phylogenetic tree using identical
methods to the ones for the 16 rRNA tree described above . The new metric is the
contrast between two signatures x and y given by

$$Co(x,y) := average_k \{c_k\} / std_k \{c_k\}, \text{ where } c_k := average_i \{| x_k - y_i |\}.$$

This new phylogenetic distance $Co(x,y)$ between genomes measures the degree of
similarity between two organisms by the contrast $\{c_k\}$ (normalized) signatures, given
by the average value of the Signal-to-Noise Ratio (SNR) in the entries in the contrast
matrix of their genomic signatures x and y The results are shown in Fig. 2 Based on
this new contrast metric, a corresponding phylogenetic tree **16 cDNA** was obtained as
shown in Fig. 3 (right). As can be observed in the differences (right column), the 16
cDNA tree is identical to the original 16S sRNA tree described above, except for a

CORRELATIONS \ EUCLIDEAN DISTANCES

	SIGNATURES					CONTRASTS				
Escherichia coli K12	15.9185	14.2954	25.4862	20.1236	33.8266	44.3278	42.3086	64.2237	57.1247	76.0675
Escherichia coli CFT073	0.8932	15.5671	23.4017	20.5336	32.3024	0.8989	42.4801	52.749	53.0186	65.1561
Salmonella enterica Choleraesuis ch	0.9054	0.8973	24.6629	20.5395	33.2218	0.9078	0.9021	60.9798	57.0622	73.2329
Photobacterium profundum SS9	0.7492	0.7881	0.7656	22.9978	34.398	0.7795	0.8157	0.7919	54.1966	59.2584
Vibrio fischeri ES114 ch1	0.8134	0.8186	0.8096	0.796	35.8779	0.8259	0.8333	0.8211	0.8223	74.5232
Pseudomonas aeruginosa PAO1	0.656	0.677	0.6662	0.621	0.5937	0.7008	0.7228	0.7079	0.6921	0.6544

Fig. 2. Correlations (lower triangles) and Euclidean distances (upper triangles) between six
bacterial signatures (column 1) on the DNA chip described in Fig. 1. They give rise to a con-
trast-based phylogenetic tree **T16 cDNA** that essentially reproduces the corresponding fragment
of an expanded generally accepted tree **16S rRNA.**

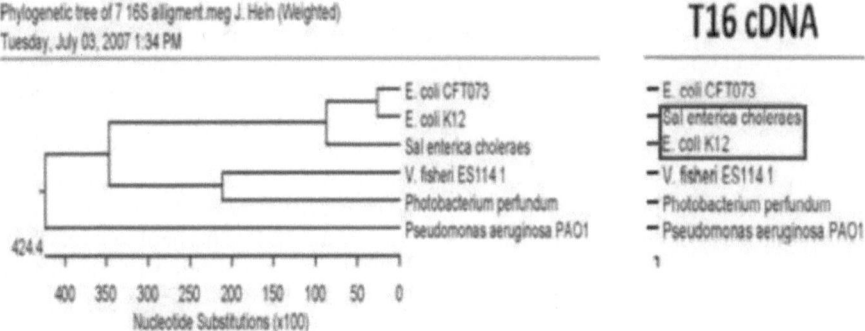

Fig. 3. 16S rRNA (left): Phylogenetic tree of six (6) bacterial chromosomes shown in Fig. 1.
The tree **T16 cDNA** (right) is identical to it, except for one difference (transposition of *Sal.
Enterica and E-Coli K12*), which is thus *essentially reproduced ab initio* by comparative genomic analyses of their (digital) genomic signatures (shown in Fig. 2) on a DNA chip of non-
crosshybridizing (nxh) 16-mer probes. The DNA chip was obtained by the PCR Selection
protocol (in simulation) from the full set of fragments of the ORFs in their genomic DNA.

permutation of two contiguous genomes. It is remarkable that this result has been obtained purely on genomic analysis, independently of the many other considerations that led to the original tree of life 16 sRNA

An additional consequence is worth pointing out. These results also provides evidence to counter the superficial observation that genomic signatures cannot provide an adequate tool for genetic comparison since they ignore the transcription process altogether (and so do not involve protein or protein expression directly.) 16S rRNA is essentially protein-based and yet, the phylogenetic tree obtained by the contrast metric from Fig. 2   shows that it is well approximated by analyses of genomic signatures alone on a 16-mer nxh DNA chip built *a priori* and designed for the collection of 6 subfamilies of plasmids and bacteria used in this validation.

4 Summary and Future Work

This paper continues the exploration of noncrosshybrizing sets developed for DNA Computing (Garzon et al., 2009; Deaton et al., 2006). as a DNA-based indexing system for biological applications. We have shown that the recently developed technique of DNA indexing (Bobba et al, 2006; Garzon et al., 2005; Garzon et al., 2004b) can be successfully applied to infer *ab initio* phylogenetic trees that are solidly established and well accepted in biological phylogenesis. The major goal of this research is to develop a methodology for enabling arbitrary species classification based on so-called universal DNA chips that cover the entire spectrum of organisms, *known or unknown.* These DNA chips can be regarded as generalized barcodes. Although barcodes are, at best, expected to discriminate among species, they do not provide an insight into the more general and complex relationships among genomes captured by phylogentic relationships, or their complex intra- or inter-genomic relationships, let alone a basis for comparative gene wide analysis.

The new DNA-based technique for genomic identification offers several other advantages. First, it is much more effectively in terms of cost and time than by traditional methods using traditional sophisticated bioinformatic analysis. For example, in another application not reported here, similar work may produce a discriminating set of probes to rapidly identify three most common types of candida strains (*C. albicans*, , *C. glabrata*, and *C. dublinensis*) in hospital infections using easily available technology. By contrast, the standard laboratory test for *Candida* involves microscopic confirmation of the fungi in the sample, followed by growth of the fungi in broth and then on solid media, followed by phenotypic identification. This method is very time-consuming (days, weeks, or even months) (Liu et al, 2005), particularly for slowly growing species, and requires mycology for accurate identification of the different species. Second, it can be applied with newly available universal DNA chips readily available both *in vitro* by the PCR Selection protocol (Deaton et al., 2006; Chen et al., 2006) and *in silico* (Garzon et al., 2009; Garzon et al., 2006) through computer simulations in virtual test tubes (Garzon et al., 2004). This means, in particular, that they have the potential to provide a universal coordinate system to characterize very large groups (families, genera, and even phyla) of organisms on a common reference system, a veritable comprehensive "Atlas of Life", as it is or as it could be on earth.

References

1. Adleman, L.: Molecular computation of solutions of combinatorial problems. Science 266, 1021–1024 (1994)
2. Blain, D., Garzon, M.H., Shin, S.Y., Zhang, B.T., Kashiwamura, S., Yamamoto, M., Kameda, A., Ohuchi, A.: Development, Evaluation and Benchmarking of Simulation Software for Biomolecule-based Computing. J. of Natural Computing 3(4), 427–442 (2004)
3. Bi, H., Chen, J., Deaton, R., Garzon, M., Rubin, H., Wood, D.H.: A PCR Protocol for in Vitro Selection of Non-Crosshybridizing Oligonucleotides. J. of Natural Computing 2(3), 417–426 (2003)
4. Bobba, K.C., Neel, A.J., Phan, V., Garzon, M.H.: "Reasoning" and "Talking" DNA: Can DNA understand english? In: Mao, C., Yokomori, T. (eds.) DNA12. LNCS, vol. 4287, pp. 337–349. Springer, Heidelberg (2006)
5. Chen, J., Deaton, R., Garzon, M., Wood, D.H., Bi, H., Carpenter, D., Wang, Y.Z.: Characterization of Non-Crosshybridizing DNA Oligonucleotides Manufactured in vitro. J. of Natural Computing 5(2), 165–181 (2006)
6. DasGupta, K.M., Konwar, I.I., Shvartsman, A.A.: Highly scalable algorithms for robust string barcoding. Int. J. Bioinformatics Research and Applications 1(2) (2005)
7. Deaton, J., Chen, J., Garzon, M., Wood, D.H.: Test Tube Selection of Large Independent Sets of DNA Oligonucleotides R, pp. 152–166. World Publishing Co., Singapore (2006); (Volume dedicated to Ned Seeman on occasion of his 60th birthday)
8. Garzon, M.H., Phan, V., Neel, A.: Optimal Codes for Computing and Self-Assembly. Int. J. of Nanotechnology and Molecular Computing 1, 1–17 (2009)
9. Garzon, M.H., Yan, H. (eds.): DNA 2007. LNCS, vol. 4848. Springer, Heidelberg (2008)
10. Garzon, M.H., Phan, V., Roy, S., Neel, A.J.: In Search of Optimal Codes for DNA-Computing. In: Mao, C., Yokomori, T. (eds.) DNA12. LNCS, vol. 4287, pp. 143–156. Springer, Heidelberg (2006)
11. Garzon, M.H., Bobba, K., Phan, V., Kontham, R.: Sensitivity and Capacity of Microarray Encodings. In: Carbone, A., Pierce, N.A. (eds.) DNA 2005. LNCS, vol. 3892, pp. 81–95. Springer, Heidelberg (2006)
12. Garzon, M.H., Blain, D., Neel, A.J.: Virtual Test Tubes for Biomolecular Computing. J. of Natural Computing 3(4), 461–477 (2004)
13. Garzon, M., Bobba, K., Hyde, B.: Digital information encoding on DNA. In: Jonoska, N., Păun, G., Rozenberg, G. (eds.) Aspects of Molecular Computing. LNCS, vol. 2950, pp. 152–166. Springer, Heidelberg (2003)
14. Hennig, W.: Grundzüge einer Theorie der Phylogenetischen Systematik English revision, Phylogenetic Systematics (tr. D. Davis and R. Zangerl). Univ. of Illinois Press, Urbana 1966, reprinted 1979 (1950)
15. Henz, S.R., Huson, D.H., Auch, A.F., Nieselt-Struwe, K., Schuster, S.C.: Whole-genome prokaryotic phylogeny. Bioinformatics 21(10), 2329–2335 (2005)
16. Liu, T.T., Lee, R.E.B., Barker, K.S., Lee, R.E., Wei, L., Homayouni, R., Rogers, P.D.: Genome-wide expression profiling of the response to azole, polyene, echinocandin, and pyrimidine antifungal agents in Candida albicans. Antimicrobial Agents and Chemotherapy 49(6), 2226–2236 (2005)
17. Margulis, L.: Symbiosis in Cell Evolution, 2nd edn. Freeman, New York (1993)
18. Neel, A.J., Garzon, M.H.: DNA-based Memories: A Survey. Studies in Computational Intelligence, vol. 113, pp. 259–275. Springer, Heidelberg (2008)

19. Neel, A., Garzon, M.: Semantic Retrieval in DNA-Based Memories with Gibbs Energy Models. Biotechnology Progress 22(1), 86–90 (2006)
20. Qiu, Q., Mukre, P., Bishop, M., Bruns, D., Wu, Q.: Hardware Accelerator for Thermodynamic Constrained DNA Code Generation. In: Garzon, Y. (ed.), pp. 201–210 (2008)
21. Reif, J.H., LaBean, T.M., Pirrung, M., Rana, V.S., Guo, B., Kingsfor, C., Wickman, G.S.: Experimental Construction of Very-Large Scale Databases with Associative Search Capability. In: Jonoska, N., Seeman, N.C. (eds.) DNA 2001. LNCS, vol. 2340, p. 231. Springer, Heidelberg (2002)
22. Seeman, N.: DNA in a material world. Nature 421, 427–431 (2003)
23. Stekel, D.: Microarray Bioinformatics. Cambridge University Press, Cambridge (2003)
24. Tulpan, D., Andronescu, M., Chang, S.B., Shortreed, M.R., Condon, A., Hoos, H.H., Smith, L.M.: Thermodynamically based DNA strand design. Nucleic Acids Res. 33(15), 4951–4964 (2005)
25. Watkins, N.E., SantaLucia Jr., J.: Nearest-neighbor thermodynamics of deoxyinosine pairs in DNA duplexes. Nucleic Acids Research 33(19), 6258–6267 (2005)
26. Winfree, E., Liu, F., Wenzler, L.A., Seeman, N.C.: Design and self-assembly of two-dimensional DNA crystals. Nature 394, 539–544 (1998)
27. Woese, C., Fox, G.: Phylogenetic structure of the prokaryotic domain: the primary kingdoms. Proc. Natl. Acad. Sci. USA 74, 5088–5090 (1977)
28. Wong, T.Y., Fernandes, S., Sankhon, N., Leong, P.P., Kuo, J., Liu, J.K.: On the role of premature stop codons in bacterial evolution. J. Bacteriology (in press, 2009)
29. Zhou, F., Olman, V., Xu, Y.: Barcodes for Genomes and Applications. Bioinformatics 9, 546 (2008)

Renewable, Time-Responsive DNA Logic Gates for Scalable Digital Circuits

Ashish Goel[1] and Morteza Ibrahimi[2]

[1] Departments of Management Science and Engineering, and by courtesy, Computer Science,
Stanford University
ashishg@stanford.edu
[2] Departments of Electrical Engineering, Stanford University
ibrahimi@stanford.edu

Abstract. In this article, we introduce a design of DNA logic gates based on enzymatic restriction of DNA strands. We present a construction for a set of one and two-input logic gates and argue that our construction can be generalized to implement any Boolean operation. A key feature of our design is its time-responsiveness, in the presence of appropriate fuels these gates can operate continuously and generate a time-dependent output in response to a time-dependant input. They can be interconnected to form digital circuits. Moreover, modulo connectivity information, the strand design and circuit design phases are decoupled.

1 Introduction

Over the last decade, several experimental breakthroughs have been obtained in the in-vitro design of circuits which operate at the molecular scale and are largely composed of (and driven by) DNA and RNA, e.g., [1], [2], [3], [4]. These approaches use circuit-elements (i.e., gates) that get consumed during the operation of these elements, restricting the applicability of these elements to feed-forward, use-once circuits. In this paper, we propose and analyze a design for molecular circuits using renewable, time-responsive, and scalable DNA gates. Our design uses a small fixed number of restriction enzymes, and uses a small fixed number of distinct fuel molecules to replenish all the gates in a circuit. The key idea of our design is to acheive a modular architecture – the computational part of the gates depends only on the logic functionality being implemented, the signaling molecules consist of two parts, a connection identifier that depends only on the location of the gate within the network, and a payload which depends only on the data encoded. In addition to being useful from efficiency considerations, renewability is also qualitatively important in that it permits the design of feedback circuits such as random access memories. We will describe our results more concretely after describing related work and establishing some notation. We would like to note that our design and analysis works in an abstract model; we do not have any experimental evidence of the efficacy of our design.

Starting from a seminal result of Adleman [5], many theoretical designs have been proposed for DNA automata and Turing machines [6,7,8,9,10]. There has also been considerable progress on the experimental side, often focused on the implementation of molecular circuits and gates that mimic their digital counterparts (e.g., [1,2,11,4,3]).

R. Deaton and A. Suyama (Eds.): DNA 15, LNCS 5877, pp. 67–77, 2009.

Stojanovic et. al. [1] presented a set of deoxyribozyme-based logic gates: NOT, AND, and XOR. As the input and output of the gates are both DNA strands, different gates can communicate with each other. Seelig et. al. [2] reported the design and experimental implementation of DNA-based digital logic gates that can be combined to construct large, reliable circuits. In addition to logic gates they demonstrated signal restoration and amplification. Their gates use single stranded DNA as inputs and outputs, and the mechanism relies exclusively on sequence recognition and strand displacement. Qian and Winfree [3] presented a set of catalytic logic gates suitable for scaling up to large circuits and developed a formalism for representing and analyzing circuits based on these gates. Macdonald et al [11] reported the development of a solution-phase logic circuit comprising over 100 molecular deoxyribozyme-based logic gates.

Engineered nucleic acid logic switches based on hybridization and conformational changes have also been successfully demonstrated in vivo [12,13]. These switches have been extended to more complex logical gates by Win and Smolke[4,14]. Their gates are part of a single molecule of RNA which can fold on itself into a special structure. It can detect specific chemical molecules as input and either cleave itself or remain intact based on the input(s) and the function of the gate. On a related note, recent advances have also been made in designing simple molecular machines that open and close like a clamp [15], walk along a track [16,17], or crawl a surface [18].

1.1   Our Results

Both the enzymatic (eg. [1]), and sequence based (eg. [3]) designs described above are non-renewable, in the sense that the functional elements (i.e. gates) of these circuits can be used only once. Removing this restriction is a major open problem, and is the main topic of study of this paper. We present an enzymatic design for DNA gates that can implement any Boolean operation. In particular the family of gates, AND, OR, NOT, and buffer can be constructed. The inputs and outputs of these DNA gates are themselves DNA strands. Furthermore, these gates operate continuously in the presence of required fuel strands and energy sources for the enzymes. Operation of these gates requires only generic fuel strands, i.e., independent of the underlying circuit, and therefore can fully emulate an electronic circuit. Our model consists of a set of DNA strands floating in a solution. A set of enzymes are also present and the solution meets the conditions for their operation. The input is introduced by adding a specific DNA strand and the output can be read from another DNA strand with the same structure as the input. These DNA strands consist of DNA fragments with constant number of bases. The length of these fragments grows logarithmically with the number of gates in the circuit.

We analyzed the stochastic behavior of these gates, as a single unit and in a cascade, in an idealized setting. Our analysis shows that the error probability for a cascade simplifies to the product of error rate and interaction rate which can be represented in a closed form. Our analysis inductively computes the Laplace transform of the error-probability and hence applies only to feed-forward circuits.

Digital circuits have played a crucial role in the technological revolution of the past decades. To a great extent, they owe this triumph to their robustness, modularity, and existence of tools for systematic design and analysis. This is also becoming an important goal in molecular circuits design. The circuits that we propose have a modular structure.

To obtain and explain this structure, we introduce a set of concepts that we incorporated in our design, and which we believe will provide a useful framework for future work in this area. We define four attributes for a DNA strand: *data obliviousness, function obliviousness, position obliviousness*, and *persistancy*. A DNA strand is data oblivious if it does not contain any information about the state of the system, i.e., it does not depend on any input value and no output value is dependent on it. A strand is function oblivious if it is not specific to any Boolean operation. A strand is position oblivious if it contains no position information, i.e., it does not correspond to any specific location in the circuit. And finally, a strand is persistance if it is present as a non-waste strand during the operation of the circuit.

Furthermore, we defien the following three classes of DNA complexes (typically two single DNA strands partially hybridized with each other) with regard to a DNA-based digital circuit: *messengers* $\mathcal{M}$ which communicate information between gates (i.e. acts as inputs/outputs), *gates* $\mathcal{G}$ which generate outputs based on the inputs, and *fuels* $\mathcal{F}$ that facilitate the manipulation of the messengers by the gates as follows:

- $D \in \mathcal{M}$ if D is function oblivious and persistance.
- $D \in \mathcal{G}$ if D is data oblivious and persistance
- $D \in \mathcal{F}$ iff D is position oblivious.

We say a design is scalable if the union of these three sets is the set of all DNA strands in the circuit. This requirement ensures that circuit design is decoupled from strand design and the circuit operation is independent of the circuit design. Although we do not prove this claim formally, the intuition is as follows. Imagine a DNA complex in a circuit that function as a gate. It cannot be position oblivious since a gate needs to be assigned to a specific position in the circuit. It cannot be function oblivious since a gate needs to implements a specific function. Hence, if the above requirement is met, it must be in the set $\mathcal{G}$ and therefore not consumed or altered and be data oblivious. Since this strand cannot be consumed or altered, it cannot function as a fuel, and since it is data oblivious it cannot function as a messenger. Similarly a messenger cannot function as a gate or fuel and a fuel cannot function as a gate or messenger. Therefore, different functionalities have to be decoupled to met the above requirement, resulting in generic fuel and decoupled circuit and strand design.

We also elaborate on a technique that we used in our design and which we believe can be applicable in other settings. the first technique *arity-conversion*, enables us to construct multi-input single-output binary logic gates using single-input single-output quaternary motifs. In electronic circuits, allowing an element to accept multiple inputs is easy, since the electrical signal can be easily confined and conducted. In contrast, DNA has a linear structure, making multiple inputs cumbersome. On the flip side, due to the analog nature of an electrical signal, performing digital operations require a mapping from the analog domain into the digital domain. In contrast, due to the combinatorial nature of DNA strands, it is straight forward to construct distinct signal strands corresponding to different states of a (potentially non-binary) digital variable without the need for any analog to digital mapping. Hence, performing digital operation is very natural even in non-binary regimes. Arity-conversion allows us to address the signal localization difficulty in DNA circuits using their above-described advantage.

It would be interesting to either duplicate our results (renewability, time-responsiveness, and scalability) using strand displacement techniques, i.e., enzyme free, or prove that it is impossible. While our current analysis applies only to feed-forward circuits, our design can also be used to obtain circuits with feedback (and hence allow latches, which in turn allow elements such as random access memory cells). It would be interesting to study the temporal and equilibrium behavior of our design in terms of error-probability in a feedback circuit.

Section 2 establishes the notation which we then use in section 3 to describe our design. Section 4 presents a brief description of our stochastic analysis results. We conclude with some open problems and a brief discussion.

2   Notation

Suppose S is a sequence of letters A, C, T, G. We denote by $\underline{S}$ a single stranded DNA molecule with sequence S starting from the 5' end. And by $\bar{S}$ a single stranded DNA molecule with the complementary sequence of S starting from the 3' end. With an slight misuse of notation we use S to also represent the double stranded DNA formed by the hybridization of $\underline{S}$ and $\bar{S}$. We denote by S^r the sequence obtained by reversing S. Similarly S^c represents the sequence obtained by taking the complement of S. $\underline{S}^r$, $\bar{S}^r$, $\underline{S}^c$, and $\bar{S}^c$ are defined as before for the sequences S^r and S^c respectively.

As a matter of convention, we use $R(e)$ to represent the recognition site of a restriction enzyme e with a single recognition site. , we use $R(\tilde{e})$ corresponding to an enzyme with a distributed recognition site that cuts in two symmetric restriction sites on both sides of the recognition site. Similarly, $R(\hat{e})$ corresponds to an enzyme with a distributed recognition site which cuts on one side or in the middle of the recognition region.

3   Single-Input and Multi-Input Boolean Operators

Our design consists of three sets of DNA strands: gate strands, messenger strands, and fuel strands. Each gate can perform a specific Boolean operation, i.e., it receives one or multiple Boolean variables as input and generates a single Boolean variable as output. Each gate also has a single physical input site (sticky end) and a single physical output site (sticky end). The sequence of each sticky end function as the corresponding input/output site's *connection ID*. Multiple input and output sites can have the same connection ID, i.e., sticky ends with the same sequence . We call the set of all the input and output sites that share the same connection ID a *connection point*. Every messenger strand also has a single connection ID encoded in the sequence of its one sticky end. However, the encoding of the connection ID in the messenger strands uses the sequence complementary to the sequence used for encoding the same connection ID in a gate. Thus, messenger strands can attach to the input and output sites in their connection point using base-pairing of their sticky ends. Therefore, the messenger strands effectively *wire* all the input and output sites in a connection point.

In addition to the connection ID, a messenger strand has a separate region called the *payload* that encodes data by assuming one of its multiple states, $\{0, 1\}$ for a binary messenger and $\{0, 1, 2, 3\}$ for a quaternary messenger. The state of the payload is determined by its sequence. Each time a messenger interacts with a gate as an output its state, i.e., encoded data, will be updated according to the gate's input(s) and function. A messenger can drive other gates, as an input, zero or more times in the interval between successive updates of its state. We say a connection point is in equilibrium if the relative concentration of messenger strands in different states at that connection point is stable.

Fuel strands are necessary to facilitate the interactions between messenger and gate strands and restore the messenger strands to their original structure after this interaction.

We start by presenting our design for the single-input single-output (SISO) gates and binary messenger strands in section 3.1. We shall then proceed by extending them into multi-input single-output (MISO) gates and quaternary messenger strands in section 3.2.

3.1   Single-Input Single-Output Gates

A non-trivial deterministic SISO Boolean operator (one in which the output depends on the input) can have its output equal to the complement of its input, *NOT* gate, or equal to its input, which we call a *translator* gate using the terminology of [2], since it translates the connection ID of the signal. For illustration purposes, in all of the strands below, assume that a letter represents a segment of some fixed length l which we call one *unit*. Let $\tilde{o}$, d, and c be restriction enzymes which cut a DNA strand as below:

$$R(\tilde{o})N^6 R(\tilde{o})^r N^8{}_\uparrow N^\downarrow$$
$$R(c)N^6{}_\uparrow N^\downarrow \qquad\qquad R(d)N^8{}_\uparrow N^\downarrow \qquad\qquad (1)$$

where $^\downarrow N_\uparrow$ indicates a restriction site, R a recognition site, and N^k an arbitrary sequence of DNA of length k units. Enzyme d is used to detach the input and output messengers from the gate. Enzyme c is used in the operation of the fuel strands. We use the enzyme $\tilde{o}$ to expose a specific sticky end on the output messenger. The exposer of this sticky end starts a sequence of reactions between the output messenger and the fuel strands which results in the output messenger being updated to the desired state, e.g., in the case of binary logic there are two different sticky ends corresponding to the output being zero or one. The proceeding paragraph describes the mechanism through which one of the sticky ends of the output messenger is exposed.

Recall that the enzyme $\tilde{o}$ has a distributed recognition site with two segments at a specific distance from each other. The first part of $\tilde{o}$'s recognition site , $R(\tilde{o})$, is placed in the messenger strand. Every gate includes two instances of the second part of $\tilde{o}$'s recognition site, $R(\tilde{o})^r$. If the first instance of $R(\tilde{o})^r$ is completed, i.e., $R(\tilde{o})$ appears at the required distance in the input messenger strand, the enzyme $\tilde{o}$ can cut the output messenger and expose the sticky end corresponding to a *one* (T_1 below). However, if the second instance of $R(\tilde{o})^r$ is completed the cut will expose the sticky end corresponding to a *zero* (T_0 below).

Using this technique, we encode the truth table for the desired Boolean operations in the position of different instances of $R(\tilde{o})$ in a messenger strand. Below we present a detailed description of the operation of these gates.

We denote by $G(I_1, I_2)$ a gate with input and output connection IDs I_1 and I_2 respectively. $M(b, I)$ represents a messenger strand with state b and connection ID I.

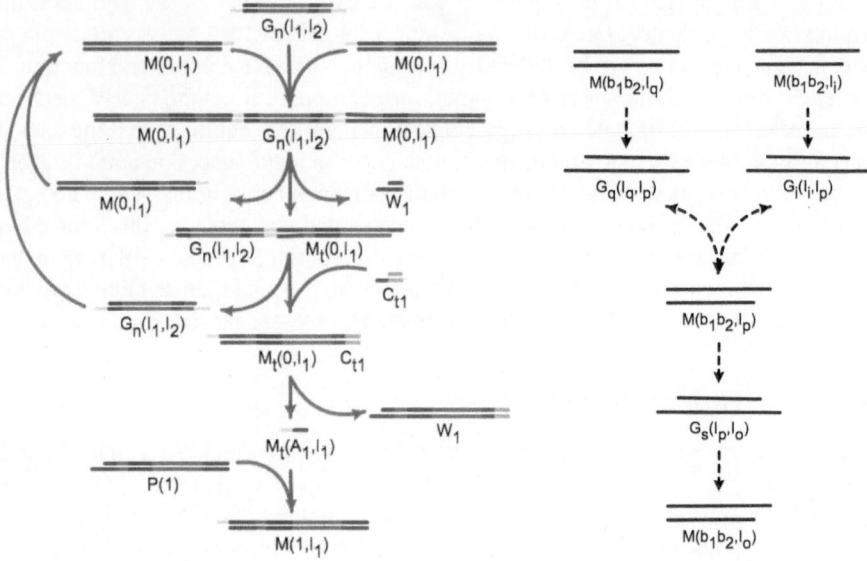

Fig. 1. Desired reaction chain of a SISO NOT gate

Fig. 2. The two-stage architecture of a MISO gate

The strands $G_n(I_1, I_2)$, $G_t(I_1, I_2)$, $M(0, I)$ and $M(1, I)$ below are the *NOT* gate, *Translator* gate, and binary messengers representing zero and one.

$$
\begin{aligned}
G_n(I_1, I_2) &: \underline{I_1} R(\tilde{o})^{rc} R(\tilde{o})^{rc} \bar{I}_2^r && M(0, I) : R(d) T_0 T_1 R(\tilde{o}) N^2 R(\tilde{o}) A_0 A_1 \bar{I} \\
G_t(I_1, I_2) &: \underline{I_1} N^2 R(\tilde{o})^{rc} R(\tilde{o})^{rc} \bar{I}_2^r && M(1, I) : R(d) T_0 T_1 N R(\tilde{o}) R(\tilde{o}) N A_0 A_1 \bar{I}
\end{aligned}
\tag{2}
$$

In addition to the gates and messenger strands, the following fuel strands are necessary for the operation of the system.

$$
\begin{aligned}
P(0) &: \quad R(d) T_0 T_1 N^2 R(\tilde{o}) N^2 R(\tilde{o}) \bar{A}_0 && C(0) : \quad R(c) \bar{T}_0^r \\
P(1) &: \quad R(d) T_0 T_1 N R(\tilde{o}) R(\tilde{o}) N A_0 \bar{A}_1 && C(1) : \quad R(c) \bar{T}_1^r
\end{aligned}
\tag{3}
$$

$P(i)$ is the payload used to reconstruct the messenger with the state i. And $C(i)$ is for translating an exposed sticky end. Consider a *NOT* gate, $G_n(I_1, I_2)$. Figure 3.1 depicts the sequence of the desired reactions for an input $M_(0, I_1)$. Assume the messenger $M_(0, I_1)$ attaches to it using the sticky end I_1 as an input and another messenger, $M(0, I_2)$, attaches using the sticky end I_2 as an output. A messenger can detach from the gate at any time via enzyme d. While both input and output messengers are attached to the gate, the following sequence of reactions can occur.

– The second instance of $R(\tilde{o})$ (from the left) in $M(0, I_1)$ and the first instance of $R(\tilde{o})^r$ in $G_n(I_1, I_2)$ constitute a valid recognition site for the enzyme $\tilde{o}$. Enzyme $\tilde{o}$ can cut the output messenger into two parts. the first part that detaches from the gate-messenger complex and floats in the solution as waste. The second part that is still attached to the gate and now has an exposed sticky end T_1. The input messenger can detach from the complex after this step without interrupting the following reactions.

- The fuel strand $C(1)$ (equation 3) can then attach to the second part of the output messenger and cut it in a way that exposes sticky end $\bar{A}_1^r$.
- Fuel strand $P(1)$ can attach using this exposed sticky end $\bar{A}_1^r$ and reconstruct the output messenger. The output messenger is now updated to a $M(1, I_2)$, the output of a *NOT* gate to a zero input.
- The enzyme d can act and detach the output messenger from the gate.

If the input to the *NOT* gate was $M(1, I_1)$ instead, similar reactions would follow. The second $R(\tilde{o})$ in $M(1, I_1)$ and the second $R(\tilde{o})^r$ constitute a valid recognition site for the enzyme $\tilde{o}$. This results in the consecutive exposure of the sticky ends T_0 and A_0 and attachment of strand $P(0)$. The output messenger can now detach, being transformed into $M(0, I_2)$. The operation of the translator gate is similar.

Observe that during the updating process two waste strands are released into the solution. While the waste strands in our system cannot cause any incorrect computation, their accumulation over time results in the inefficiency of the system in terms of time and energy.

3.2   Multi-Input Single-Output Gates

As before, the connectivity of different elements is defined through matching sticky ends. Each DNA strand can have two sticky ends in a straight forward way. However, for a MISO gate at least three connections are required. To address this issue we use a simple but potentially powerful technique for constructing a MISO gate from the SISO gates of the previous section. The core idea is that a quaternary variable can carry as much information as two binary variables. Therefore, a gate with two binary inputs can be emulated by a single-input gate which accepts quaternary inputs. We only discuss two-input gates here, it is clear that this scheme can be generalized to multi-input gates.

Based on this idea, a two-input gate can be implemented in two stages. In the first stage the two binary inputs are translated into a single quaternary signal. The second stage performs the desired Boolean operation on the quaternary signal. Figure 1 depicts this architecture. Both these stages can be implemented using the ideas of the previous section with minor modification.

Above all, the messenger strand should be modified to encode two bits of information, i.e., take four different states. The set of restriction enzymes is also needed to be modified for the operation of the new system. Let $\tilde{o}$, $\hat{d}$, and c be restriction enzymes with distributed recognition sites which cut a DNA strand as below:

$$R(\tilde{o})N^{12}R(\tilde{o})^r N^{16}{}_\uparrow N^\downarrow$$
$$R(c)N^{14}{}_\uparrow N^\downarrow \qquad\qquad R(d)N^{16}{}_\uparrow N^\downarrow \qquad\qquad (4)$$

Messenger Strand: Following the same idea as in the SISO gates, the messenger strand is modified as below:

$A(I) : A_{00}A_{10}A_{01}A_{11}\bar{I}$

$Q(00) : NR(\tilde{o})NR(\tilde{o})R(\tilde{o})N^2R(\tilde{o})$ $Q(01) : NR(\tilde{o})R(\tilde{o})N^2R(\tilde{o})R(\tilde{o})N$

$Q(10) : NR(\tilde{o})R(\tilde{o})NR(\tilde{o})N^2R(\tilde{o})$ $Q(11) : R(\tilde{o})NR(\tilde{o})N^2R(\tilde{o})R(\tilde{o})N$

$T(00) : T_{00}^2 T_{10}^2 T_{00}^1 T_{01}^1$ $T(01) : T_{01}^2 T_{11}^2 T_{00}^1 T_{01}^1$

$T(10) : T_{00}^2 T_{10}^2 T_{10}^1 T_{11}^1$ $T(11) : T_{01}^2 T_{11}^2 T_{10}^1 T_{11}^1$

$M(ij, I) : R(d)T(ij)Q(ij)A(I)$

The messenger strand can now take four different forms, i.e., it encodes two bits of information. The current state of a messenger, $M(ij, I)$ can be inferred from its T segment. T_{is}^1 is a sticky end whose exposure results in the first bit of the messenger to be updated to the value s, updating the messenger to the state is. Therefore, the current value of the second bit is i. Similarly, T_{rj}^2 represents a messenger with the current value of the first bit equal to j.

Similar to the previous section, the messenger strand consists of two separate parts. The first part, $A(I)$, contains the connection information of the messenger. This segment is independent of the data encoded by the messenger and is common among all messenger strands with the same connection ID. The second part, $T(ij)Q(ij)$, does not depend on the connection ID and encodes only the data. Hence, this part is common among all the messengers which encode the same data. Moreover, these two parts can be separated by a single cut. This enables us to use a simple mechanism to update the data encoded in a messenger strand using location oblivious fuel, i.e., independent of the connectivity mapping of the messenger and the underlying circuit.

First Stage: In our two-stage implementation of a two-input gate, the first stage consists of two modified translator SISO gates $G_i(I_i, I_p)$ and $G_q(I_q, I_p)$ .Both $G_i(I_i, I_p)$ and $G_q(I_q, I_p)$ read only the first bit of their quaternary input. But they write on different bits of their common output messenger, call it $M(b_2b_1, I_p)$. Gate $G_i(I_i, I_p)$ writes the least significant bit (LSB) of its input on the most significant bit (MSB) of $M(b_2b_1, I_p)$, i.e., b_2, while gate $G_q(I_q, I_p)$ writes the LSB of its input on the MSB of $M(b_2b_1, I_p)$, i.e., b_2. Therefore, at equilibrium, $M(b_2b_1, I_p)$ has its MSB equal to the first input, i.e., input to $G_i(I_i, I_p)$, and its LSB equal to the second input, i.e., input to $G_q(I_q, I_p)$.

The structures of $G_i(I_i, I_p)$ and $G_q(I_q, I_p)$ are essentially the same as the SISO gate introduced in the previous section with a slight difference in their length which results in cutting the output in different locations, i.e., T^2 and T^1 for gates $G_i(I_i, I_p)$ and $G_q(I_q, I_p)$ respectively.

$$G_i(I_i, I_p) : \underline{I_i}N^6R(\tilde{o})^rR(\tilde{o})^r\bar{I_p} \qquad G_q(I_q, I_p) : \underline{I_q}N^6R(\tilde{o})^rR(\tilde{o})^rN^2\bar{I_p}$$

Second Stage: The second stage consists of a SISO gate which can respond to a quaternary signal. Below are the gate strands for the *AND*, and *OR* operations. We have also included the *NOT* and translator gates which are modified to function with the quaternary messenger strands. Note that these two gates are single stage SISO gates and depends only on the LSB of a quaternary messenger strand as their input.

$$\begin{aligned} G_{sa}(I_p, I_o) &: \underline{I_p}R(\tilde{o})^rR(\tilde{o})^r\bar{I_o} & G_{so}(I_p, I_o) &: \underline{I_p}N^2R(\tilde{o})^rR(\tilde{o})^r\bar{I_o} \\ G_n(I_p, I_o) &: \underline{I_p}N^4R(\tilde{o})^rR(\tilde{o})^r\bar{I_o} & G_t(I_p, I_o) &: \underline{I_i}N^6R(\tilde{o})^rR(\tilde{o})^r\bar{I_p} \end{aligned} \qquad (5)$$

Fuel Strands: In addition to the messenger and gate, the following fuel strands are required.

$$P(00) : R(d)T(00)Q(00)\bar{A}_{00} \qquad\qquad P(01) : R(d)T(01)Q(01)A_{00}\bar{A}_{01}$$
$$P(10) : R(d)T(10)Q(10)A_{00}A_{01}\bar{A}_{10} \qquad P(11) : R(d)T(11)Q(11)A_{00}A_{01}A_{10}\bar{A}_{11}$$
$$C_{t(i,j)} : R_cN^{13-l(i,j,k)}\bar{T}_{ij}^k \quad i,j,k \in \{0,1\}$$

$$(6)$$

where $l(i, j, k)$ is the number of segments between T_{ij}^k and A_{ij} in the messenger strand.

The strand $C_{t(i,j)}$ can attach to an exposed $\underline{T}_{ij}^r$ sticky end of a messenger and consequently make a cut on the A_{ij} part of the messenger exposing a $\underline{A}_{ij}^r$ sticky end. The strands $P(ij)$ are payload strands that can attach to an exposed $\underline{A}_{ij}^r$ sticky end of a messenger and restore it to the state ij.

The operation of a MISO gate follows the same steps as the SISO gates. We do not present a detailed description here due to limited space.

4 Stochastic Analysis of the Gate's Output

It has been mentioned that the messenger strands at the output of a gate move to an equilibrium state depending on the input(s) and the function of the gate. In this section we investigate the dynamics of this process by asking how fast the output approaches equilibrium and what this equilibrium is. We answer these questions by considering the relative concentration of the updated messengers as a stochastic process in time and characterizing its first order distribution. We only present the results of this analysis here.

We assume that after a messenger strand attaches to a gate as an output it will either be updated to a new state with some time-dependent probability, or remain intact. the probability of a messenger strand being updated to an incorrect state is also time-dependent. Further, we assume that a messenger strand spends a negligible amount of time in a gate-messenger complex. We also assume that the time for a messenger strand to attach to a gate strand has exponential distribution with parameter λ where λ is a function of the gate and the messenger concentrations and other conditions of the solution. Given these two assumptions, the number of times a messenger strand attaches to any gate in an interval $(0, t]$, m, has a Poisson distribution with parameter λt. For any messenger strand, let $x(t)$ be the indicator function for the correctness of the current state of the messenger strand, i.e., $x(t) = 1$ if messenger strand is in the desired state at time t and $x(t) = 0$ otherwise. Consider a messenger strand $M(b_2^i b_1^i, I_o)$ which is the output messenger for a gate $G(I_i, I_o)$. Our goal in this section is to calculate the probability of the messenger strand $M(b_2 b_1, I_o)$ being in the desired state $M(b_2^f b_1^f, I_o)$ at time t, i.e., $P\{x(t) = 1\}$.

Consider a cascade of SISO gates, $G(I_0, I_1), G(I_1, I_2), G(I_2, I_3), ..., G(I_{n-1}, I_n)$ and number them from one to n. Let $x_j(t)$ be defined as before for the messengers $M(b_1 b_2, I_j)$, i.e., the messengers that communicate between gates $j - 1$ and j. Also assume that λ_j, as defined before, is the effective rate of the Poisson process according to which messenger $M(b_1 b_2, I_j)$ is being updated.

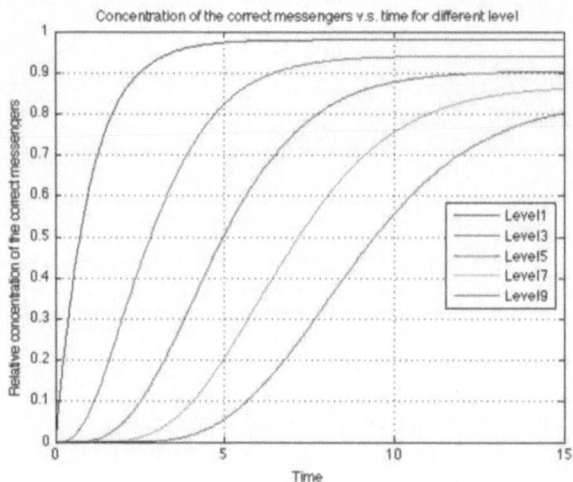

Fig. 3. Concentration of the correct messengers v.s. time for different stage

We showed that $E[x_j(t)]$ can be found using:

$$X_j(s) = \prod_{i=1}^{j} \frac{\zeta}{\lambda_i + s} X_0(s) \qquad (7)$$

where $X_j(s)$ is the Laplace transform of the expected relative concentration of the correct messenger strands at stage j, $\mathcal{L}\{E[x_j(t)]\}$.

Given an input to the first stage in the cascade $E[x_j(t)]$ can be calculated using Eq. 7. For example, for an input applied at time $t = 0$ which does not change with time, i.e., $x_0(t) = \zeta^{-1} q_0 u(t)$ with $u(t)$ the step function, and for $\lambda_1 = \lambda_2 = ... = \lambda_n = \lambda$ we will get $E[x_j(t)] = q_0 \zeta^{j-1} \gamma(\lambda t, j)$. where $\gamma(x, j)$ is the upper gamma function. Figure 3 shows the expected relative concentration of the correct messenger strands at stage j.

5   Conclusion and Open Problems

Our goal in this paper was to develop a set of logic gates that can be used to construct scalable, renewable, and time-responsive digital circuits. Moreover, we required a modular design, one in which the design of DNA strands would be as independent of the underlying circuit as possible. We developed a set of concepts that formalize these requirements and used them as our guideline in designing the system. We also performed a stochastic analysis of the error-probability of our design.

Our design is based on enzymatic restriction of DNA strands. While we use a small number of different enzymes in our design, namely four, a non-enzymatic implementation with the above properties or a formal proof of impossibility would be of interest.

While our analysis only works for feed-forward circuits, our design can also be used to obtain circuits with feedback (and hence allow latches, which in turn allow elements such as random access memory cells). It would be interesting to study the temporal and equilibrium behavior of our design in a feedback circuit.

References

1. Stojanovic, M., Mitchell, T., Stefanovic, D.: Deoxyribozyme-Based Logic Gates. Journal-American Chemical Society 124(14), 3555–3561 (2002)
2. Seelig, G., Soloveichik, D., Zhang, D., Winfree, E.: Enzyme-Free Nucleic Acid Logic Circuits (2006)
3. Qian, L., Winfree, E.: A simple DNA gate motif for synthesizing large-scale circuits. In: Proceedings of the 14th International meeting on DNA based computers (2008)
4. Win, M., Smolke, C.: From the Cover: A modular and extensible RNA-based gene-regulatory platform for engineering cellular function. Proceedings of the National Academy of Sciences 104(36), 14283 (2007)
5. Adleman, L.: Molecular computation of solutions to combinatorial problems. Science 266, 1021–1024 (1994)
6. Rothemund, P.: A DNA and restriction enzyme implementation of Turing machines. In: DNA Based Computers: Proceedings of a DIMACS Workshop, April 4, 1995. Princeton University, American Mathematical Society (1996)
7. Benenson, Y., Gil, B., Ben-Dor, U., Adar, R., Shapiro, E.: An autonomous molecular computer for logical control of gene expression. Nature 429(6990), 423–429 (2004)
8. Yin, P., Sahu, S., Turberfield, A., Reif, J.: Design of Autonomous DNA Cellular Automata. In: Carbone, A., Pierce, N.A. (eds.) DNA 2005. LNCS, vol. 3892, pp. 399–416. Springer, Heidelberg (2006)
9. Chen, H., Anindya, D., Goel, A.: Towards Programmable Molecular Machines. In: Proceeding of 5th Conference on Foundation of Nanoscience, pp. 137–139 (2008)
10. Yin, P., Turberfield, A., Sahu, S., Reif, J.: Design of an autonomous DNA nanomechanical device capable of universal computation and universal translational motion. In: Ferretti, C., Mauri, G., Zandron, C. (eds.) DNA 2004. LNCS, vol. 3384, pp. 426–444. Springer, Heidelberg (2005)
11. Macdonald, J., Li, Y., Sutovic, M., Lederman, H., Pendri, K., Lu, W., Andrews, B., Stefanovic, D., Stojanovic, M.: Medium scale integration of molecular logic gates in an automaton. Nano. Lett. 6, 2598–2603 (2006)
12. Isaacs, F., Dwyer, D., Ding, C., Pervouchine, D., Cantor, C., Collins, J.: Engineered riboregulators enable post-transcriptional control of gene expression. Nature Biotechnology 22, 841–847 (2004)
13. Bayer, T., Smolke, C.: Programmable ligand-controlled riboregulators of eukaryotic gene expression. Nature Biotechnology 23, 337–343 (2005)
14. Win, M., Smolke, C.: Higher-Order Cellular Information Processing with Synthetic RNA Devices. Science 322(5900), 456 (2008)
15. Yurke, B., Turberfield Jr., A.M., Simmel, F., Neumann, J.: A DNA-fuelled molecular machine made of DNA. Nature 406, 605–608 (2000)
16. Yin, P., Yan, H., Daniell, X.G., Turberfield, A.J., Reif, J.H.: A unidirectional DNA walker moving autonomously along a linear track. Angew. Chem. Int. Ed. 43, 4906–4911 (2004)
17. Sherman, W., Seeman, N.: A precisely controlled DNA bipedal walking device. Nano. Letters 4, 1203–1207 (2004)
18. Shin, J.S., Pierce, N.: A synthetic DNA walker for molecular transport. Journal of American Chemistry Society 126, 10834–10835 (2004)

Self-assembly of the Discrete Sierpinski Carpet and Related Fractals

Steven M. Kautz and James I. Lathrop

Iowa State University, Ames, IA 50011 U.S.A.
smkautz@cs.iastate.edu, jil@cs.iastate.edu

Abstract. It is well known that the discrete Sierpinski triangle can be defined as the nonzero residues modulo 2 of Pascal's triangle, and that from this definition one can construct a tileset with which the discrete Sierpinski triangle self-assembles in Winfree's tile assembly model. In this paper we introduce an infinite class of discrete self-similar fractals (a class that includes both the Sierpinski triangle and the Sierpinski carpet) that are defined by the residues modulo a prime p of the entries in a two-dimensional matrix obtained from a simple recursive equation. We prove that every fractal in this class self-assembles and that there is a uniform procedure that generates the corresponding tilesets. As a special case we show that the discrete Sierpinski carpet self-assembles using a set of 30 tiles.

1 Introduction

A model for self-assembly is a computing paradigm in which many small components interact locally, without external direction, to assemble themselves into a larger structure. Wang [8,9]. first investigated the self-assembly of patterns in the plane from a finite set of square tiles. In Wang's model, a tile is a square with a label on each edge that determines which other tiles in the set can lie adjacent to it in the final structure. The *Tile Assembly Model* of Winfree [10], later revised by Rothemund and Winfree [6,5], refines the Wang model to provide an abstraction for the physical self-assembly of DNA molecules.

We introduce some formal notation for the Tile Assembly Model in the next section. Briefly, however, a tile can be viewed as a square that has a string on each edge, called the *color*, along with an integer *binding strength* which in this paper is always 0, 1, or 2. A tile may also have a label for informational purposes. Two tiles can lie adjacent to each other only if the adjacent edges have the same color and the same binding strength. Intuitively, the binding strength and edge color model the potential strength of a bond between corresponding "sticky ends" of specially constructed DNA molecules.

A tile system is assumed to start with an infinite supply of a finite number of tile types. A set of initial tiles, the seed assembly, is placed in the discrete plane. Self-assembly proceeds nondeterministically as new tiles bind to the existing assembly. In this process, tiles may cooperate to create a planar structure, or (by appropriate interpretation of the labels) perform a computation. Winfree

R. Deaton and A. Suyama (Eds.): DNA 15, LNCS 5877, pp. 78–87, 2009.

[10] and others [6,5,2,1] have shown that such systems can perform computations such as counting and addition, and that in fact the model is universal: given an arbitrary Turing machine, there is a tile set for which each row of the resulting assembly is the result of a computation step of the Turing machine. It is also possible to use a finite tile set to generate infinite planar structures such as discrete fractals. An example of the latter was made famous when Papadakis, Rothemund and Winfree [4] performed an experiment in which DNA molecules were used to self-assemble a portion of the discrete Sierpinski triangle.

The discrete Sierpinski triangle has been used extensively as a test structure for in DNA self-assembly [10]. One reason for this is that it self-assembles using a simple set of only 7 tiles. More generally, however, fractal structures are of interest because "[s]tructures that self-assemble in naturally occurring biological systems are often fractals of low dimension, by which we mean that they are usefully modeled as fractals and that their fractal dimensions are less than the dimension of the space or surface that they occupy. The advantages of such fractal geometries for materials transport, heat exchange, information processing, and robustness imply that structures engineered by nanoscale self-assembly in the near future will also often be fractals of low dimension [2]." It is then natural to ask what discrete fractals there are, other than the ubiquitous Sierpinski triangle, that can self-assemble with a relatively small set of tile types in this model.

In this paper we introduce an infinite class of self-similar discrete fractals, all of which self-assemble in Winfree's model. The class includes, as special cases, the standard Sierpinski triangle and Sierpinski carpet. All the fractals in this class exhibit a strong self-similarity property that we call *numerical self-similarity*. Each fractal is defined in terms of an infinite integer matrix M whose entries are residues modulo a given prime q. A fractal S, viewed in the usual way as a subset of the discrete plane, can then be defined as the set of points (i,j) for which $M[i,j]$ is not congruent to zero, modulo q. The usual notion of self-similarity for a set S means that for some integer p, S is completely determined by membership in S of the points (s,t) with $0 \leq s,t < p$; given any $k \geq 0$ and $i,j < p^k$, the point (sp^k+i, tp^k+j) is in S if and only if $(i,j) \in S$ and $(s,t) \in S$. That is, the set of points in the p^k by p^k square whose lower left corner is at (sp^k, tp^k) is either empty, if $(s,t) \notin S$, or is an exact copy of the p^k by p^k square with lower left corner at the origin, if (s,t) is in S.

Numerical self-similarity means further that the entries of the p^k by p^k submatrix of M with lower left corner at $M[sp^k, tp^k]$ are always exactly related to those of the p^k by p^k submatrix at the origin by a factor of $M[s,t]$, that is, $M[sp^k + i, tp^k + j] \equiv M[s,t]M[i,j]$ for $i,j < p^k$. We show that there exists a simple recursively generated matrix M that defines the discrete Sierpinski carpet using $p = q = 3$. Figure 1 is a depiction of the mod 3 residues of the matrix M. It follows from Theorem 3, Theorem 1, and the simple definition of the matrix M, that the Sierpinski carpet self-assembles in the Tile Assembly Model.

The next section introduces some definitions and notation for the Tile Assembly Model described above. Section 3 reviews some known results on the

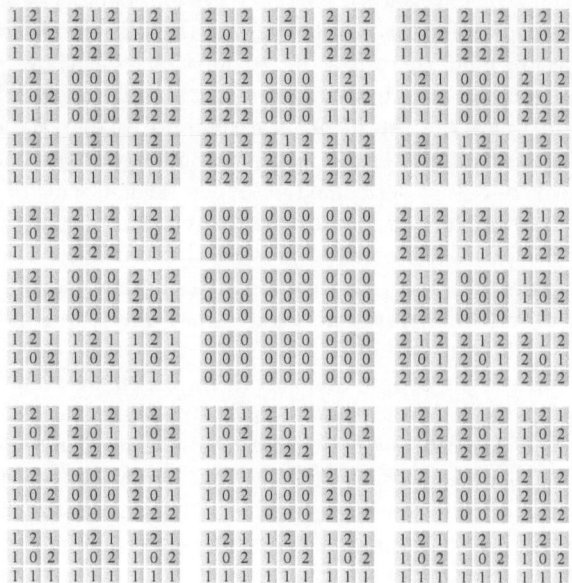

Fig. 1. First three stages of the numerically self-similar Sierpinski carpet

relationship between the Sierpinski triangle and Pascal's triangle and describes a uniform construction of tilesets for recursively defined matrices. In Section 4 we define the matrices from which we obtain discrete fractals and prove the main result on numerical self-similarity of such matrices. Section 5 contains some concluding remarks and open problems.

2   Preliminaries

In this section we briefly introduce some notation and terminology associated with the Tile Assembly Model described in the introduction. For full details see [6,5,10].

We work in the discrete Euclidean plane. Let Σ be a finite alphabet. A *tile type* t is a mapping that associates a *color* $\mathrm{col}_t(\boldsymbol{u}) \in \Sigma^*$ and a *strength* $\mathrm{str}_t(\boldsymbol{u}) \in \mathbb{N}$ with each of the four sides of a unit square, where the side is indicated by the unit vector $\boldsymbol{u}$. We also assume that there is a *label* associated with each tile type by a function $m : T \to \mathcal{L}$, where $\mathcal{L}$ is a finite alphabet (which in the examples of interest will be a set $\{0, 1, \ldots, p-1\}$ for some prime p).

Let T denote a finite set of tile types and let $\tau \in \mathbb{N}$ be a fixed parameter, called the *temperature* (which in the present paper is always 2). We assume that there is an infinite supply of tiles for each type $t \in T$. A tile may be positioned, but not rotated, in the discrete plane. Two adjacent tiles may *bond* if the abutting edges have matching color and matching strength s; the strength of the bond is

s. More generally, an *assembly* is a partial assignment $\alpha : \mathbb{Z} \to T$ of tile types to locations in the plane in which each tile is bonded to its neighbors with a *total* strength of at least τ, and such that the assembly cannot be separated into smaller assemblies without breaking a set of bonds having a total strength of at least τ.

The process of self-assembly begins with a given *seed assembly* σ and proceeds nondeterministically by extending the domain of the assembly, where a new tile may extend an assembly at position (x, y) if all edges abutting those of existing adjacent tiles have matching colors and matching strengths and if the sum of the strengths for the abutting edges is at least τ. An assembly is *terminal* if it cannot be extended. A *tile assembly system (TAS)* is a triple (T, σ, τ) where T is a finite set of tile types, σ is the seed assembly, and τ is the temperature. A TAS is *definitive* if it has a unique terminal assembly.

Definition 1. *Let $\mathcal{L}$ be a finite alphabet, let I denote a finite or infinite subinterval of $\mathbb{N}$, and let $M : I^2 \to \mathcal{L}$ be a matrix. M self-assembles if there exists a definitive TAS (T, σ, τ) with terminal assembly α, and a labeling $m : T \to \mathcal{L}$, such that $\alpha(x, y)$ is defined wherever $M[x, y]$ is defined and for all (x, y) in the domain of M, $m(\alpha(x, y)) = M[x, y]$. We may say that a subset $S \subseteq I^2$ self-assembles if there is a subset $\mathcal{L}'$ of $\mathcal{L}$ such that $(x, y) \in S \iff M[x, y] \in \mathcal{L}'$.*

The following useful fact is not difficult (although somewhat tedious) to formalize and prove. The proof gives a uniform procedure for constructing a tileset from a given finite function f.

Theorem 1. *For any matrix M, if there is a positive integer n such that each entry $M[x, y]$ is determined by a finite function f of the entries $M[x', y']$ with $x - n < x' \leq x$ and $y - n < y' \leq y$ (excluding $M[x, y]$ itself), then M self-assembles.*

3   Tiling the Sierpinski Triangle

In this section we formally introduce the notion of numerical self-similarity mentioned in the introduction and review some known results regarding self-assembly of the Sierpinski triangle. The following definition generalizes the usual definition of a discrete self-similar fractal.

Definition 2. *Let $p, q \geq 2$. Let $M : \mathbb{N}^2 \to \mathbb{N}$. M is numerically p-self-similar modulo q if for all $0 \leq s, t < p$, for all $k \geq 0$ and for all $i, j < p^k$,*
$M[sp^k + i, tp^k + j] \equiv M[s, t] \cdot M[i, j] \pmod{q}$.

Suppose a matrix M is defined on $\mathbb{N}^2$. It is not difficult to see that if M is numerically p-self-similar modulo q and if $S \subseteq \mathbb{N}^2$ is the set defined such that $(x, y) \in S \iff M[x, y] \not\equiv 0 \pmod{q}$, then S is a discrete self-similar fractal in the usual sense.

We define Pascal's triangle to be the matrix P of integers defined for all $i, j \geq 0$ by $P[i, 0] = 1$, $P[0, j] = 1$, and $P[i, j] = P[i, j-1] + P[j, i-1]$ for $i, j > 0$.

Thus $P[i,j] = \binom{i+j}{j}$, where the reverse diagonal $i+j = k$ corresponds to the usual kth row of Pascal's triangle rendered horizontally. The relationhip between Pascal's triangle and the discrete Sierpinski triangle is well known; the set S defined by $(x,y) \in S \iff P[x,y] \equiv 0 \pmod 2$ is the Sierpinski triangle, a fact implicit in Winfree's construction of a 7-tile system with which the Sierpinski triangle self-assembles [10]. This fact is an instance of a more general result.

Theorem 2. *For any prime p, P is numerically p-self-similar modulo p.*

Theorem 2 has been known in various forms for some time, e.g., a proof can be found in [3], but it is also a special case of of our Theorem 3, which is proved using different techniques.

4   Tiling the Sierpinski Carpet

In this section we introduce a class of matrices defined by a simple recursion and prove our main technical result, which states simply that these matrices are numerically self-similar. As a consequence, we are able to conclude in Corollary 1 that the Sierpinski carpet self-assembles.

Definition 3. *Let a, b, $c \geq 0$. Let M be defined for all $i, j, \geq 0$ by*

$$M[0,0] = 1$$
$$M[0,j] = a^j \ for \ j > 0$$
$$M[i,0] = c^i \ for \ i > 0$$
$$M[i,j] = aM[i,j-1] + bM[i-1,j-1] + cM[i-1,j] \ for \ i,j > 0 \quad (1)$$

Theorem 3. *M is numerically p-self-similar modulo p.*

Before moving to the proof of Theorem 3 we note the following corollary. Let $a = b = c = 1$ and $p = 3$ in Definition 3, Let $S \subset \mathbb{N}$ be the set of points (x, y) such that $M[x, y] \not\equiv 0$ modulo 3. Then S is self-similar and is evidently the Sierpinski carpet. By Theorem 1, S self-assembles. Thus we have shown:

Corollary 1. *The discrete Sierpinski carpet self-assembles.*

Although we have not formally proved Theorem 1 in this paper, given the relation (1) it is not difficult to show that a set of 30 tiles is sufficient.

We next proceed to prove some preliminary results leading to the proof of Theorem 3.

When $a = b = c = 1$, the entries of M are known as the *Delannoy numbers*. The following is a very slight generalization of a well-known combinatorial interpretation of the Delannoy numbers; see the Encyclopedia of Integer Sequences A001850 [7]. Note that a similar expression holds when $i > j$.

Lemma 1. *For $i \leq j$,*

$$M[i,j] = \sum_{k=0}^{i} \binom{j}{k} \binom{j+i-k}{i-k} a^{j-k} b^k c^{i-k}. \tag{2}$$

Throughout the remainder of the discussion below, we work with a fixed prime p, fixed integers a, b, and c, and a matrix M defined as in Definition 3. Since all arithmetic will be modulo p, we can assume the entries of M are the residues modulo p.

Definition 4. *Let $M(x, y, u)$ denote the finite $u \times u$ submatrix of M whose lower left corner is at (x, y); that is, $M(x, y, u)[i, j] = M[x + i, y + j]$ for $0 \leq i < u$, $0 \leq j < u$.*

Definition 5. *Let $k \geq 0$. A 1-block of size p^k is the matrix $M(0, 0, p^k)$. For $0 \leq n < p$, an n-block of size p^k is a $p^k \times p^k$ matrix B for which $B \equiv nM(0, 0, p^k)$.*

Observation 1. Note that in the terms of the preceding definitions, M is numerically p-self-similar modulo p if and only if for all $s, t < p$ and all $k \geq 0$, $M(sp^k, tp^k, p^k)$ is a $M[s, t]$-block of size p^k. Note also that an n-block satisfies (1) modulo p, and therefore is completely determined by its first row and first column.

Lemma 2. *Let $k \geq 0$. Then $M[0, p^k - 1] \equiv M[p^k - 1, 0] \equiv 1$.*

Proof. By definition, $M[0, p^k - 1] \equiv a^{p^k - 1}$ and $M[p^k - 1, 0] \equiv c^{p^k - 1}$. Using Fermat's little theorem and a simple induction on k, $a^{p^k} \equiv a \mod p$, so $a^{p^k - 1} \equiv 1$. Similarly, $c^{p^k - 1} \equiv 1$.

Lemma 3. *(a) Let $k > 0$, $t < p$, $j < p^k$, and $n = M[0, t] = a^t$. Then $M(0, tp^k, p^k)[0, j] \equiv na^j$.*
(b) Let $k > 0$, $s < p$, $i < p^k$, and $n = M[s, 0] = c^s$. Then $M(sp^k, 0, p^k)[i, 0] \equiv nc^i$.

Proof. By definition, $M(0, tp^k, p^k)[0, j] = M[0, tp^k + j] \equiv a^{tp^k + j} \equiv (a^{p^k})^t a^j$, where the latter is equivalent to $a^t a^j$ using Fermat's little theorem as in the proof of the previous lemma. The second part is similar.

Lemma 4. *For all $0 < i, j < p$,*

(a) $aM[i-1, p-1] + bM[i, p-1] \equiv 0$.
(b) $bM[p-1, j-1] + cM[p-1, j] \equiv 0$.

Proof. Note that for $0 < k < p$, $\binom{p}{k}$ is divisible by p, so for $i < p$,

$$M[i, p] \equiv \sum_{k=0}^{i} \binom{p}{k} \binom{p+i-k}{i-k} a^{p-k} b^k c^{i-k} \equiv \binom{p}{0} a^p c^i \equiv a^p c^i.$$

Then by definition, for $0 < i < p$,

$$aM[i, p-1] + bM[i-1, p-1] \equiv M[i, p] - cM[i-1, p]$$
$$\equiv a^p c^i - c(a^p c^{i-1})$$
$$\equiv 0.$$

The argument for (b) is similar.

Lemma 5. (a) Let $0 \le x < y$ and fix a row $i > 0$. Let $n = M[i, x]$. Suppose that for each column between x and y, the sum of adjacent entries of row $i - 1$, when scaled by the coefficients b and c, is 0 modulo p; that is, for all $0 < j < y - x$,

$$bM[i-1, x+j-1] + cM[i-1, x+j] \equiv 0.$$

Then $M[i, x+j] \equiv na^j$ for all $0 \le j < y - x$.
(b) Let $0 \le u < v$ and fix a column $j > 0$. Let $n = M[u, j]$. Suppose that for rows between u and v, the sum of adjacent entries of column $j - 1$, when scaled by the coefficients a and b, is 0 modulo p; that is, for all $0 < i < v - u$,

$$aM[u+i, j-1] + bM[u+i-1, j-1] \equiv 0.$$

Then $M[u+i, j] \equiv nc^i$ for all $0 \le i < v - u$.

Proof. (a) For $j = 0$ we have $M[i, x+0] = n$. Having shown for an induction that $M[i, x+j] \equiv na^j$,

$$M[i, x+j+1] \equiv aM[i, x+j] + bM[i-1, x+j] + cM[i-1, x+j+1] \quad (3)$$
$$\equiv aM[i, x+j] \quad (4)$$
$$\equiv na^{j+1}. \quad (5)$$

The proof for (b) is similar.

Proof of Theorem 3. We establish (a)–(d) below by induction on k. Note that part (a) implies that M is numerically p-self-similar.

(a) For $s, t < p$ and $n = M[s, t]$, $M(sp^{k-1}, tp^{k-1}, p^{k-1}) \equiv n \cdot M(0, 0, p^{k-1})$. That is, $M(sp^{k-1}, tp^{k-1}, p^{k-1})$ is an n-block of size p^{k-1}.
(b) Let $n < p$ and Let B be any n-block of size p^k. Then for all $0 < i, j < p^k$,

$$aB[i, p^k-1] + bB[i-1, p^k-1] \equiv 0 \text{ and} \quad (6)$$
$$bB[p^k-1, j-1] + cB[p^k-1, j] \equiv 0. \quad (7)$$

(c) Let $n < p$ and Let B be any n-block of size p^k. Then

$$B[0, p^k-1] \equiv B[p^k-1, 0] \equiv n \text{ and} \quad (8)$$
$$B[p^k-1, p^k-1] \equiv n. \quad (9)$$

(d) Let $s, t < p$ and let

$$A = M((s-1)p^k, tp^k, p^k),$$
$$B = M((s-1)p^k, (t-1)p^k, p^k),$$
$$C = M((s-1)p^k, tp^k, p^k), \text{ and}$$
$$D = M(sp^k, tp^k, p^k).$$

Suppose that A is an x-block, B is a y-block, and C is a z-block and let $w \equiv ax + by + cz$. Then D is a w-block.

Note first that (d) follows from (b) and (c). That is, fix k and assume that (b) and (c) hold, and assume the hypothesis of (d). Then applying (c),

$$A[0, p^k - 1] = x, \tag{10}$$
$$B[p^k - 1, p^k - 1] = y, \text{ and} \tag{11}$$
$$C[p^k - 1, 0] = z, \tag{12}$$

and so $D[0,0] = w$. By (b), the rightmost column of A satisfies the hypothesis of Lemma 5(b), so $D[i,0] = wc^i$ for $i < p^k$. A similar argument shows that $D[0,j] = wa^j$ for $j < p^k$. Thus $D[i,0] = wM[i,0]$ and $D[0,j] = wM[0,j]$ for $i, j < p^k$, so we have $D[i,j] = wM[i,j]$ for all $i, j, < p^k$ by Observation 1; hence D is a w-block.

We next establish the base step for $k = 1$. Part (a) asserts only that $M[s,t] = M[s,t] \cdot M[0,0]$. Part (b) follows from Lemma 4. For part (c), (8) is immediate from Lemma 3 and (9) follows from (b) using the observation that for $0 \le i < p$,

$$M[i, p-1] \equiv \begin{cases} 1 & \text{if } i \text{ is even} \\ p-1 & \text{if } i \text{ is odd} \end{cases}$$

(and if $p = 2$, $p - 1 = 1$). Part (d) follows in general from (b) and (c) as shown above.

Now let $k \ge 1$ and assume the induction hypothesis holds. We first show that for $t < p$, $M(0, tp^k, p^k)$ is an n-block of size p^k, where $n = M[0,t]$. The $t = 0$ case is the definition of a 1-block. Suppose we have shown that $C = M(0, (t-1)p^k, p^k)$ is an m-block, where $m = M[0, t-1]$. Let $D = M(0, tp^k, p^k)$. We know from Lemma 3 that $D[0,j] \equiv na^j$ for $j < p^k$ and in particular $D[0,0] \equiv n$. Since C is an m-block of size p^k, by part (b) of the induction hypothesis the rightmost column of C satisfies the hypothesis of Lemma 5(b), and hence $D[i,0] \equiv nc^i$ for all $i < p^k$. Since $D[i,0] = nM[i,0]$ for $i < p^k$ and $D[0,j] = nM[0,j]$ for $j < p^k$, we have $D[i,j] = nM[i,j]$ for all $i, j < p^k$ using Observation 1. Thus D is an n-block. A similar argument shows that $M(sp^k, 0, p^k)$ is an $M[s,0]$-block for each $s < p^k$.

The next step is to show, by an induction on s and t, that $M(sp^k, tp^k, p^k)$ is an $M[s,t]$-block for all $s, t < p$. Let $x = M[s, t-1]$, $y = M[s-1, t-1]$, $z = M[s-1, t]$, and $w = M[s,t] = ax + by + cz$. and suppose that

$$A = M(sp^k, (t-1)p^k, p^k) \text{ is an } x\text{-block},$$
$$B = M((s-1)p^k, (t-1)p^k, p^k) \text{ is a } y\text{-block, and}$$
$$C = M((s-1)p^k, tp^k, p^k) \text{ is a } z\text{-block}.$$

Then by part (d) of the induction hypothesis, $M(sp^k, tp^k, p^k)$ is a $M[s,t]$-block. This establishes part (a).

Next, having established (a) we know that $B = M(sp^k, (p-1)p^k, p^k)$ is an $M[s, p-1]$-block for each $s < p$. Using part (b) of the induction hypothesis, we have

$$aB[i, p^k - 1] + bB[i - 1, p^k - 1] \equiv 0 \tag{13}$$

for $0 < i < p^k$. In addition, if $s < p-1$ and $A = M((s+1)p^k, (p-1)p^k, p^k)$, then A is a $M[s+1, p-1]$-block, so by part (c) of the induction hypothesis, $A[0, p^k - 1] \equiv M[s+1, p-1]$ and $B[p^k - 1, p^k - 1] \equiv M[s, p-1]$. It follows that

$$aA[0, p^k - 1] + bB[p^k - 1, p^k - 1] \equiv 0. \tag{14}$$

Now consider any $0 < i < p^{k+1}$; then

$$aM[i, p^{k+1} - 1] + bM[i - 1, p^{k+1} - 1] \equiv 0, \tag{15}$$

which follows from (14) if i is a multiple of p^k and from (13) otherwise. Multiplying (15) by n yields (6). The argument for (7) is similar. This establishes (b).

To prove (c) note first that by Lemma 2, $M[0, p^{k+1} - 1] \equiv M[p^{k+1} - 1, 0] \equiv 1$. Then let $B = M((p-1)p^k, (p-1)p^k, p^k)$. B is a 1-block by (a), since $M[p-1, p-1] \equiv 1$, so using (9) from the induction hypothesis,

$$M[p^{k+1} - 1, p^{k+1} - 1] = B[p^k - 1, p^k - 1] \equiv 1.$$

Then (c) is obtained by multiplying the equivalences above by n. That (d) follows from (b) and (c) has already been shown above, concluding the induction. $\square$

5   Conclusion

We have shown that the discrete Sierpinski carpet self-assembles in Winfree's Tile Assembly Model and, moreover, that it is an instance of an infinite class of discrete fractals that self-assemble. The key ingredient of this result was Theorem 3, which stated that certain recursively generated infinite matrices have a strong self-similarity property, which we defined as numerical self-similarity. Theorem 3 is a significant generalization of known results on self-similarity in Pascal's triangle which underlie the previous work on self-assembly of the discrete Sierpinski triangle.

The recursively generated matrices we study provide a rich source of examples of self-similar fractals, despite the obvious simplicity of the linear function used to generate them. We have investigated matrices generated using more complex recursive relationships and the preliminary results are inconclusive; that is, although there are examples that appear to generate fractal structures (subsets of the plane with dimension strictly less than 2), none of the structures observed so far is self-similar.

The discrete fractals we have investigated all self-assemble in a "progressive" way; that is, a tile that binds at location (x, y) is always determined by tiles at locations (x', y') with $x' \leq x$ and $y' \leq y$. An interesting question is that of finding an exact characterization of the self-similar fractals that can be tiled progressively. For example, a restriction to progressive tiling rules out self-similar fractals with blocks of zeros along either axis. It remains open whether every symmetric, self-similar discrete fractal without zeros along the axes can be tiled progressively.

Acknowledgments

The authors wish to thank Jack Lutz for useful discussions.

References

1. Lathrop, J.I., Lutz, J.H., Patitz, M.J., Summers, S.M.: Computability and complexity in self-assembly. In: Beckmann, A., Dimitracopoulos, C., Löwe, B. (eds.) CiE 2008. LNCS, vol. 5028, pp. 349–358. Springer, Heidelberg (2008)
2. Lathrop, J.I., Lutz, J.H., Summers, S.M.: Strict self-assembly of discrete Sierpinski triangles. In: Cooper, S.B., Löwe, B., Sorbi, A. (eds.) CiE 2007. LNCS, vol. 4497, pp. 455–464. Springer, Heidelberg (2007)
3. Peitgen, H.O., Jürgens, H., Saupe, D.: Chaos and fractals: New frontiers of science. Springer, Heidelberg (2004)
4. Papadakis, N., Rothemund, P., Winfree, E.: Algorithmic self-assembly of dna sierpinski triangles. PLoS Biology 12 (2004)
5. Rothemund, P.W.K.: Theory and experiments in algorithmic self-assembly. Ph.D. thesis, University of Southern California (December 2001)
6. Rothemund, P.W.K., Winfree, E.: The program-size complexity of self-assembled squares (extended abstract). In: Proceedings of the Thirty-Second Annual ACM Symposium on Theory of Computing, pp. 459–468 (2000)
7. N.J. A Sloane, The on-line encyclopedia of integer sequences, [Online; accessed 09-January-2008] (2008)
8. Wang, H.: Proving theorems by pattern recognition – II. The Bell System Technical Journal XL(1), 1–41 (1961)
9. Wang, H.: Dominoes and the AEA case of the decision problem. In: Proceedings of the Symposium on Mathematical Theory of Automata, New York, pp. 23–55. Polytechnic Press of Polytechnic Inst. of Brooklyn, Brooklyn (1962/1963)
10. Winfree, E.: Algorithmic self-assembly of DNA. Ph.D. thesis, California Institute of Technology (June 1998)

Automatic Design of DNA Logic Gates Based on Kinetic Simulation

Ibuki Kawamata, Fumiaki Tanaka, and Masami Hagiya

Graduate School of Information Science and Technology, University of Tokyo
7-3-1 Hongo, Bunkyo-ku, Tokyo, 113-8656, Japan
ibuki@is.s.u-tokyo.ac.jp, fumiaki@dna-comp.org, hagiya@is.s.u-tokyo.ac.jp

Abstract. Recently, DNA logic gates and DNA machines have been developed using only a simple complementary base pairing of DNA, that is, hybridization and branch migration. Because such reaction systems have been designed by trial and error, it has been difficult to design a complex system and to correctly verify the reaction. The purpose of this research is to develop a method for automatically searching and designing DNA logic gates based on a kinetic simulation. Since the solution space that should be searched is quite large, a simulated-annealing method is used to search for a highly evaluated system from many candidates and find a semi-optimal one. A simulator based on a kinetic model is developed, which calculates the time change of concentrations of abstracted DNA molecules. An evaluation function, in which the evaluation value rises when the logic gate works correctly, is also designed. The effectiveness of the proposed method is evaluated experimentally with an AND gate, which is designed automatically.

1 Introduction

In recent years, molecular computing has become an important aspect of nanotechnology. DNA devices, such as enzyme-free DNA logic gates [1] and entropy-driven reactions [2] have been developed. A reaction graph [3] assists in the designing of DNA-oriented systems by representation of assembly pathways. To synthesize a larger-scale logic gate, a simple DNA gate motif, called a seesaw gate, was designed [4]. Since these devices were often designed by trial and error, a great deal of effort had been required to design such systems.

The main purpose of this work is to develop a method for designing a DNA logic gate system using computer simulation. To give one solution to this problem, we developed a searching method for automatically designing a DNA logic gate using a kinetic simulation. The searching method is an algorithm that increases the evaluation value, which indicates how correctly the gate works. This evaluation value is calculated using a simulator that estimates the time change of the concentration based on kinetics. Various logic gates limited to two inputs and one output were designed using our method. We conducted an experiment with an AND gate, and showed that our method works in that case.

R. Deaton and A. Suyama (Eds.): DNA 15, LNCS 5877, pp. 88–96, 2009.

2   Methods

Figure 1 shows the flow chart of the automatic design of a DNA logic gate system. We regard this design of a DNA logic gate as a problem of maximizing the evaluation value. Starting from a random initial state, the system increases the evaluation value gradually by exchanging the state with a neighbor state, which is a local change. Though the last state after the iteration may not be the global optimum, it is possible to design a semi-optimal system, which has the highest evaluation value in the iteration, using this algorithm. In this section, a DNA reaction simulator is discussed first. Second, our developed evaluation function using the simulator is explained. The final section describes our method that is used to search for a system with higher evaluation, which is calculated with this evaluation function.

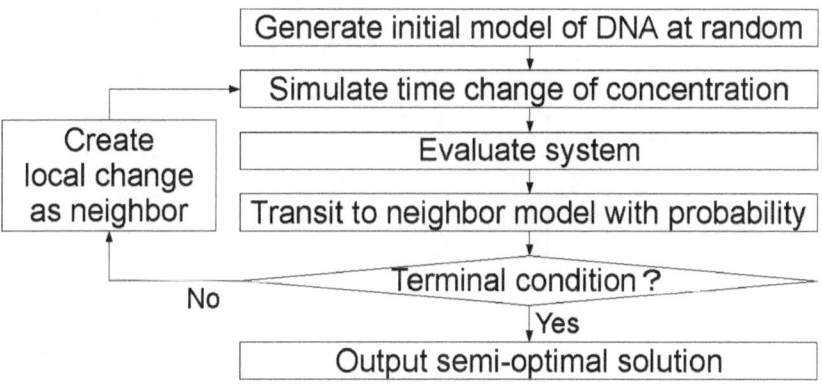

Fig. 1. Flow chart of automatic design of a DNA logic gate system

2.1   Kinetic Simulation

Simulation Model. A simulator is used to calculate the reactions of DNA molecules and the time change of the virtual concentration of each structure with an abstraction and approximation.

Model of DNA. To investigate DNA reactions on the simulator, we developed a computational DNA model. Although a single-stranded DNA molecule is a sequence of bases, the simulator does not recognize each base, and the DNA is abstracted by designating a sequence of bases as a segment. By this abstraction, a single DNA strand can be expressed by a sequence of segments. One letter of the alphabet was allocated to one segment, and uppercase and lowercase letters were used to express complementary base pairs. We treated a DNA structure as a graph, in which the nodes consist of segments and the edges consist of bonds between segments. The upper part of Figure 2 shows the model of an enzyme-free AND gate [1] and the lower part shows our abstraction.

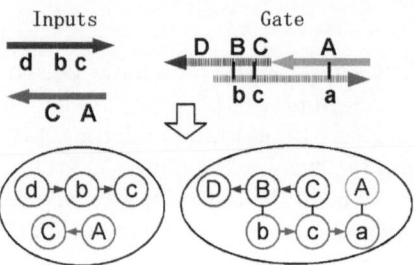

Fig. 2. Example of DNA model (enzyme-free logic gate [1])

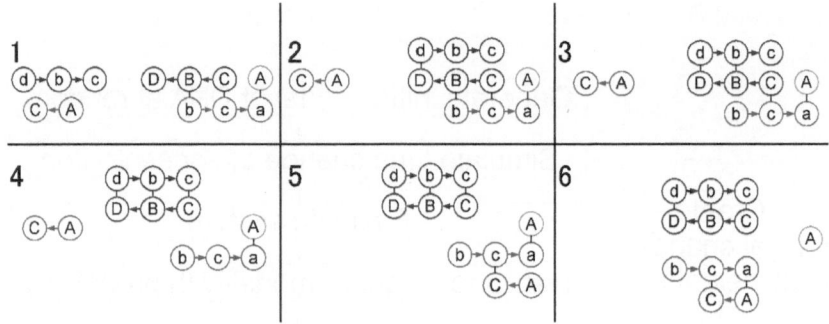

Fig. 3. Example of DNA reactions (enzyme-free logic gate [1])

To design simple DNA systems, only three reactions are simulated as interactions or internal actions of DNA strands, that is to say, hybridization, branch migration, and denaturation. In the simulator, hybridization corresponds to the addition of bond information between unconnected complementary segments, and denaturation corresponds to the opposite, which is the removal of the bond. Branch migration corresponds to exchanging base pairing information of the shorter connection to a new longer connection. Hybridization can be divided into three groups, hybridization of two different structures into one structure (type 1), intramolecular hybridization that extends the double helix by not forming a loop (type 2), and intramolecular hybridization by forming a loop (type 3). Figure 3 shows the series of graph transformations, which correspond to the reactions of the enzyme-free AND gate [1] with both inputs.

Kinetics. Chemical kinetics were used to simulate the time change of the virtual concentration of each DNA structure. For instance, the rate of change for a reaction $A + B \rightarrow C$ is calculated by the equation;

$$\frac{d[C]}{dt} = k[A][B]$$

where k is the rate constant. In the simulator, the rate constant, which is a fixed value that determines the speed of each reaction rate, was not calculated exactly and a virtual value was used. It is considerably difficult to estimate the true value of the three reactions because the rate constant depends on the base sequence. For this reason, we fixed the speeds of hybridization and branch migration at an approximate constant, and only that of denaturation was changeable as a parameter. The rate constant of hybridization type 1 is set to 0.1, type 2 to 0.2, and type 3 to 0.001 and branch migration to 0.005. The simulator distinguishes the hybridization of a segment in a loop structure because it has a special behavior. Part of a strand in a loop does not hybridize simply to another strand as a free single strand does [5]. For this reason, the speed of hybridization is modified to be fifty times slower than the speed of branch migration if either of the connected segments is in a loop. Although we need to adjust these values carefully in the future, these values are enough to design a simple logic gate.

Differential Equation. With the assumptions mentioned above, the simulation becomes a problem to solve a differential equation with an initial condition. The initial condition is given as pairs of structures and concentrations. To solve the differential equation, we used Heun's method.

A concentration threshold was introduced because the number of predicted structures would cause an explosion of combinations. In other words, most of the structures simulated were not the main product of the reaction. Structures that do not have a higher concentration than the threshold concentration 10^{-5} were disregarded. This limitation prevents this explosion of combinations, and stabilizes the simulation.

2.2   System Evaluation

A kinetic simulation is used to evaluate a DNA logic gate system, and the evaluation value indicates how correct the system works as a logic gate. The evaluation value is a weighted average of v_1 which is a gap value, and v_2, which is an average gap;

$$\text{evaluation value} = \frac{9v_1 + v_2}{10}.$$

v_1 is calculated as

$$v_1 = min(T) - max(F),$$

where T is a set of concentrations of the output with inputs that return a true logical value, and F is a set of concentrations of the output with inputs that return a false logical value. The gap value is the main point of the estimation. This calculation subtracts the worst value of the false input from the worst value of the true input. If this gap is big, the system is recognized as working correctly from the viewpoint that it is possible to distinguish the true state from the false state.

Let n be the number of total single strands in the system, S be the set of single stranded structures, and C_i be the concentrations of four states of structure i calculated with the simulation,

$$v_2 = \frac{1}{n} \sum_{s \in S} (max(C_s) - min(C_s)).$$

This function calculates the average gap of each concentration of the structure after simulation. If v_2 is high, it is possible to understand that a more complex reaction is taking place in the system being evaluated.

2.3   Design of System

Our method can be used to design a DNA logic gate that consists of three collections of strands, that is, inputs, a gate, and output. The output contains one strand that is chosen automatically by selecting from the gate, which returns the highest evaluation value.

Search Algorithm. To design a highly evaluated system that satisfies the given condition, a heuristic method was used. We used a simulated-annealing algorithm to search for a semi-optimal DNA logic gate system. The simulated-annealing algorithm gradually increases the evaluation value, and it is possible to escape from a local solution while the temperature parameter is high enough.

The outline of the algorithm is as follows, first, find a random answer for an initial state, and initialize the temperature parameter. Second, make another state with some changes as a neighbor. If the exchange condition, which is to avoid falling into a local maximum, is satisfied, then change the state to the neighbor. If the condition is not satisfied, the state remains. Next, cool the temperature parameter as scheduled. Repeat the procedure from the second step until the terminal condition. Finally output the present state as an answer.

Settings of Initial Solution and Neighbors. A random initial state is constructed from a random gate and output, and random inputs. By assuming a random strand as a sequence of random segments with a random length, random inputs consist of two random strands. A random gate is a collection of four or less strands. The random length of a strand is a number between 1 and 4, and random segments are chosen from 'a' to 'f' or from 'A' to 'F'. Each number is modifiable by changing the condition of the method. One strand is chosen from

Table 1. Probabilities of each change

change of system	change inputs	0.3
	change gate	0.7
change of strand	insert one segment	0.3
	remove one segment	0.3
	flip one segment	0.4

the gate as an output because of the assumption that an output is automatically generated from the gate.

A neighbor state is a local change of the present state, which is calculated by changing one of the inputs or gate strands, adding a random strand to the gate or removing a random strand from the gate. Table 1 shows the probabilities used to generate a random neighbor.

3   Experimental Results

By using our method, the design of six kinds of logic gates, which were AND, OR, NAND, NOR, NOT and XOR gate, was experimented. The difference among the experiments was only the T and F sets throughout the evaluation. As a result of the search, several kinds of logic gates were provided automatically, and some examples are introduced below.

Figure 4 shows an example of the topology of an AND gate system found using our search method. Input 1 "AB" displaces the output strand "A" by branch migration and wastes a structure in which "AB" and "ba" are together. Although the output strand becomes single stranded, the output strand hybridizes to another strand making a bond between 'A' and 'a'. Input 2 "AC" similarly displaces the output if the reaction of input 1 is completed. If either input is insufficient, the output strand will not be emitted. This is how this DNA logic gate system acts as an AND gate. The calculated evaluation value was about 0.92. There was a segment attached to one of the gate strands that had no function in the reaction.

A simulation was carried out to investigate the time change of the output concentration. The results of simulating the AND gate is shown in Figure 5. The concentration of the output increased if and only if both inputs were added.

To prove the correctness of our method, we experimented with the AND gate system designed by our method using the strands shown in Table 2. The AND gate was checked by the change of fluorescent intensity (Figure 6) at 37 °C using a F-2500 Fluorescence Spectrophotometer (HITACHI). The excitation and emission wavelengths for FAM were respectively 494 and 518 nm. And all oligonucleotides were first dissolved in a 10 mM Tris Buffer with a pH of 7.4

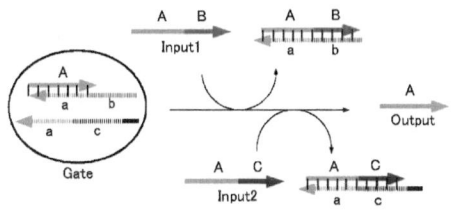

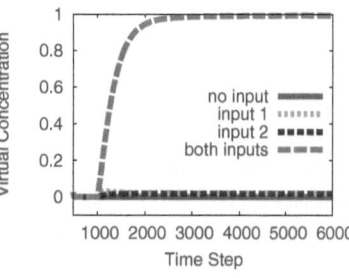

Fig. 4. Designed AND Gate

Fig. 5. Simulation of output concentration of AND gate

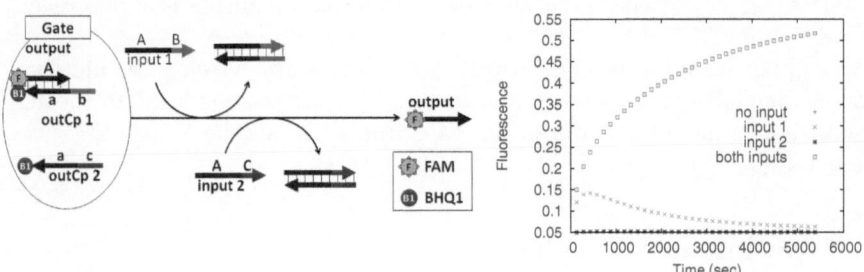

Fig. 6. Experimental AND Gate

Fig. 7. Fluorescence experiment

Table 2. Sequence of oligonucleotides (5' end to 3' end)

Input 1	CCAAACTACTTACGTTGAACATACACCGAGGTTTAGTGAACACTTCTAAGCAACTAA
Input 2	TGAACATACACCGAGGTTTAGTCCAAA
Output	TGAACATACACCGAGGTTTAG
OutComp 1	TGTTCACTAAACCTCGGTGTATGTTCA
OutComp 2	TTTGGACTAAACCTCGGTGTATGTTCA

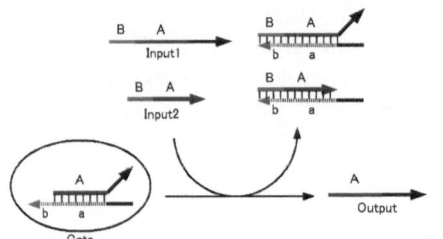

Fig. 8. Designed OR Gate

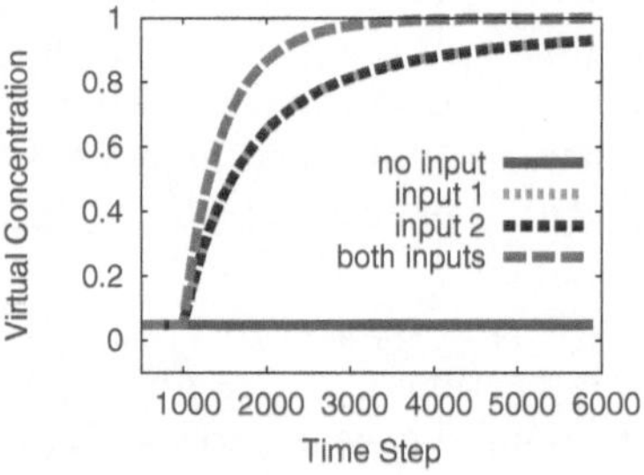

Fig. 9. Simulation of output concentration of OR gate

and then diluted with a 1 SSC buffer containing 0.15 M of NaCl and 15 mM of sodium citrate with a pH of 7.0. Figure 7 shows the change in fluorescence. It is clear that the system works correctly as an AND gate because the fluorescence is measured highly only if both inputs are added.

Although we were able to find an OR, NAND, NOR and NOT gate, we have not tested the gates in vitro. OR and NAND gates are explained below. Figure 8 is an OR gate designed by our method, where both inputs can displace the output by branch migration using segment 'b' as a toehold. This OR gate is essentially the same as the Seelig's enzyme-free OR gate [1]. Though three segments are unnecessary for the reaction, it was found that this DNA logic system functions as an OR gate, from the results of simulating the output concentration (Figure 9). Figures 10 and 11 are the same as those for the NAND gate. Input 1 is a hairpin structure with segment 'a' in a loop. Although the output is a single strand with one segment 'A', the speed of hybridization between these segments is regulated on the assumption that a segment in a loop is treated as an exception. Input 2 opens the hairpin structure of input 1 by branch migration, and segment 'a'

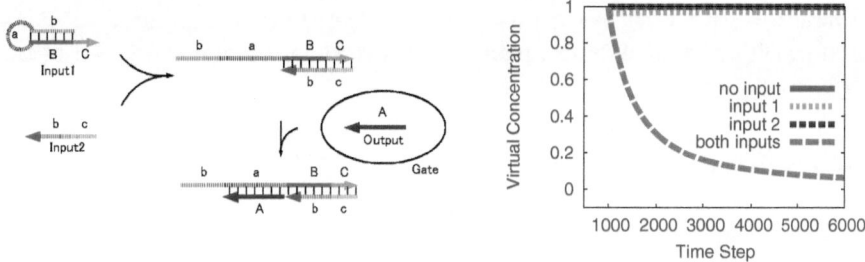

Fig. 10. Designed NAND Gate

Fig. 11. Simulation of output concentration of NAND gate

becomes accessible. As a result, the concentration of the output strand decreases by hybridizing to the segment. The time change of the output concentration is shown in Figure 11, which indicates that this DNA logic system behaves as a NAND gate.

4 Discussion

By listing the examples that our method found and testing in vitro, it appears a simple DNA logic gate can be automatically designed using this method. Among these DNA logic gate systems, branch migration and hybridization were the center of the reaction. The reason for segments or strands that have no effect on a DNA logic gate system may be that those segments do not affect the evaluation value.

The results of the fluorescence experiment and simulation are basically the same, although details of the two graphs are different. The difference was caused by not adjusting the rate constant between simulation and the fluorescence experiment. Although it was possible to design an XOR gate with a high evaluation value using our method, the gate worked well on the simulation and did not seem to be adaptable to a chemical experiment. This is because of the rough abstraction and approximation of the simulator. These results suggest that our method has potential for designing a more complex DNA logic gate system by improving the simulator.

By assuming the third type of hybridization, there is a possibility to compose a pseudoknot structure, even if it is not a natural structure. It should be noted that the simulation of DNA concentration does not correspond to the actual concentration. Although it is a virtual concentration, it seems that the simulator can calculate the behavior of the DNA strand as a whole. Two inputs in the gate found using our method are not completely different, for the inputs can have the same segment in common. Because our method was developed to design the topology of a system constructed only with DNA, it does not recognize reactions with enzyme and does not support sequence design.

Because our method searches for only logic gates, further tasks should include searching for other devices such as a DNA walker, DNA comparator, or exponential growth system. We also plan to improve the accuracy and speed of the kinetic simulation.

5   Conclusion

Since it has been a task to design a nano-device based on DNA, we developed a program for automatically designing a DNA logic gate system using a kinetic simulation. Although there were several limitations, it was possible to design several DNA logic gates using the method, and an AND gate worked correctly with the fluorescence experiment. By conquering the limitations, this system will support more applications in the future.

References

1. Seelig, G., Soloveichik, D., Zhang, D.Y., Winfree, E.: Enzyme-free nucleic acid logic circuits. Science 314(5805), 1585–1588 (2006)
2. Zhang, D.Y., Turberfield, A.J., Yurke, B., Winfree, E.: Engineering entropy-driven reactions and networks catalyzed by DNA. Science 318(5853), 1121–1125 (2007)
3. Yin, P., Choi, H.M.T., Calvert, C.R., Pierce, N.A.: Programming biomolecular self-assembly pathways. Nature 451(7176), 318–322 (2008)
4. Qian, L., Winfree, E.: A simple dna gate motif for synthesizing large-scale circuits. In: Proceedings of 14th International Meeting on DNA Computing (DNA14), pp. 139–151 (2008)
5. Seelig, G., Yurke, B., Winfree, E.: Catalyzed relaxation of a metastable DNA fuel. Journal of the American Chemical Society 128(37), 12211–12220 (2006)

Design of a Biomolecular Device That Executes Process Algebra

Urmi Majumder and John H. Reif

Department of Computer Science,
Duke University, Durham, NC, USA
{urmim,reif}@cs.duke.edu

Abstract. *Process algebras* are widely used to define the formal semantics of *concurrent communicating processes*. In this paper, we implement a particularly expressive form of process algebra, known as stochastic π-calculus, at the *molecular* scale by providing a design for a *DNA-based biomolecular device* that simulates a process algebraic machine. Our design of the molecular stochastic π-calculus system makes use of a modified form of *Whiplash-PCR* (WPCR) machines. In this design, we connect (via a tethering DNA nanostructure) a number of DNA strands, each of which corresponds to a WPCR machine. This collection of WPCR machines are used to execute distinct concurrent processes, each with its own distinct program. Furthermore, their close proximity enables computation to proceed via communication.

1 Motivation

1.1 Process Algebra

Process algebra has been popularly used to design concurrent, distributed and mobile systems [1]. Process algebra has also traditionally been seen as a paradigm for practical concurrent languages and as a specification language for software and hardware systems that are encoded in more pragmatic ways. The main idea in process algebra is to model processes as *communicating systems* with decentralized control. In other words, in this model of computation, concurrent processes can be specified to execute distinct programs as well as communicate repeatedly with another process or set of processes. Process communication can be enforced to be *synchronous* via *handshake communication* protocols that require acknowledgment of message reception. Also, processes can replicate and generate new processes. This paper considers *stochastic π-calculus*, a particularly expressive form of process algebra developed by Cardelli [2] that provides a clear specification of probabilities of process behavior such as stochastic delays, sequentialization of processes and communication.

Stochastic π-calculus has been shown to be particularly useful for modeling biological systems. Here, biological components are modeled as concurrent processes and their interaction is treated as process communication. A precise connection between process algebra and chemical reactions was established by

R. Deaton and A. Suyama (Eds.): DNA 15, LNCS 5877, pp. 97–105, 2009.
© Springer-Verlag Berlin Heidelberg 2009

Cardelli [2]. These modeling techniques provide researchers with better models and simulations of living matter. Consequently, they provide a better understanding of how nature works and, sometimes, a tool to predict unknown behavior of living systems.

Biology already has several sophisticated examples of inter-cellular communication. For instance, bacteria use such communication for gene regulation [3]. Yet another fascinating example is the interaction that takes place during the developmental transition of fertilization which initiates a rapid series of changes that restructures an egg into a zygote. There are several signaling agents that mediate several of these rapid modifications in cell structure. Studies indicate that elements from several key signaling pathways, co-localize on molecular scaffolds in the egg and provide a means for these pathways to interact [4].

1.2   The Need for a Molecular Process Algebraic System

The central question considered in this paper is how to implement process algebraic systems with biomolecules. This direction of research would allow us to use biological matter as flexible information and material processing devices. The most important feature of process algebra is that this model allows computation to proceed via interprocess communication. At the molecular scale, adding the capability to interact with each other will allow us to build far more complex molecular computing devices than those that have been proposed to date, such as nano-scale distributed computing systems or nano-scale sensing systems.

1.3   The Challenges of a Molecular Process Algebraic System

One of the primary challenges of molecular process algebra is *local execution* of *distinct programs* meaning *parallel execution* of such programs in several machines *without interference*. For this, we need to design a biomolecular device, where multiple copies of the same or distinct devices can simultaneously compute without interfering with each other. These needs can be satisfied by a number of known biomolecular computing device designs including *tiling assemblies* [5], *Whiplash PCR machines* [6] and *hybridization chain reactions* [7].

One considerably more challenging design requirement is the requirement for *handshaking communication* between two (or more) processes, where one of the interacting processes might send data to the other and wait to resume its own program execution until it receives acknowledgment from the other process. To implement synchronous process communication at the molecular scale, carrier molecules going from process A to process B are insufficient; process B must be able to send an acknowledgment back to A that allows A to resume its execution.

1.4   Implementation of a Molecular Process Algebraic System via Modified Whiplash PCR Machines

Whiplash PCR (WPCR) machines [6] appear to be an ideal candidate for a molecular implementation of process algebra. Recall that WPCR machines are

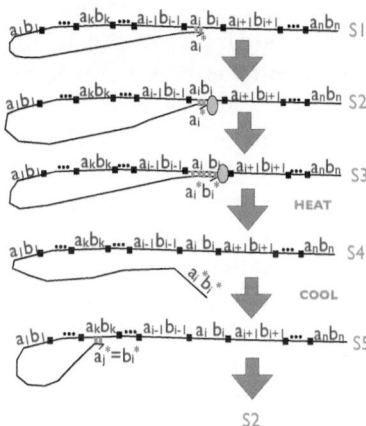

Fig. 1. Schematic of the protocol for the original Whiplash PCR machine: (S1) Initial state of the WPCR strand W with current state being a_i^*. (S2) Polymerase binds to the $3'$ end of W (holding the current state). (S3) Next state b_i^* is copied at the $3'$ end of W by primer extension. Polymerase does not copy anything beyond b_i because of the stopper sequence indicated in this figure by a black square, (S4) The mixture is heated so that W loses its stem-loop structure. (S5) The solution is cooled so that the $3'$ end of W can bind to the new current state $b_i^* = a_j^*$ encoded at the $3'$ end of the strand and the whole state transition repeats beginning with State S2.

DNA devices composed of a single strand of DNA containing a distinct program (via segments of the DNA that encode computational rewrite rules) and a current state (via a short segment at the $3'$ end of the strand). Its computational steps are executed by repeated rounds of thermal-cycling that comprises of cooling (that facilitates hybridization of the $3'$ end with an interior segment that encodes a computational rule), polymerase extension (to copy the new current state of the machine from the rewrite rule) and heating (to release the extended the $3'$ end encoding the new current state). Hence, these machines can hold both programs and their inputs in close proximity, allowing parallel computation. WPCR machines can also be connected (via a tethering DNA nanostructure) as shown in Figure 3(Left). This particular set up would allow the same two copies of process P_1 and process P_2 to communicate not only the first time but as many times needed thereafter. Consequently, a collection of WPCR machines can be used to execute distinct concurrent processes, each with their own distinct program that can communicate when required.

1.5 Biochemical Techniques Used in the Molecular Implementation of Process Algebra

The biochemical techniques that allow us to implement the various primitives of process algebra are primarily *hybridization* and *polymerization*[8]. Additionally, we

modify the original WPCR machine to support the process communication construct. These modifications include incorporation of additional secondary structure such as a *stem-loop* in each computational rewrite rule as well as in the current
state of the machine and use of a *multi-temperature thermal cycle* that starts in a
very high temperature T_0 where all the DNA strands (each representing a process)
are denatured, followed by a lower temperature T_1 $(T_0 > T_1)$ that facilitates hybridization, polymerization and restriction and back to the higher temperature T_0.
Intermittently, the temperature is cooled to T_2 $(T_1 > T_2)$ so that secondary structure such as stem-loop formation is facilitated. The stem loops allow for successful
process communication (discussed in Section 7). The selective hybridization at T_1
and T_2 can be controlled through careful design of the secondary structure (e.g.
hybridization length and sequence composition). In our implementation, polymerization uses a *non-strand-displacing polymerase* (such as Taq). The use of a non-
strand-displacing enzyme is the key to a successful process communication.

1.6   Contributions

In this paper we present a molecular implementation of process algebra using
design techniques similar to that of the original WPCR machines. We implement
the basic primitives of process algebra in this new biomolecular device (which we
call π-Whiplash PCR (π-WPCR)) machine including *stochastic delay operation,
parallel composition and process interaction*. We call the new machine π-*WPCR
machine* since we implement the stochastic π-calculus version of process algebra
at the molecular scale.

2   Process Algebra Overview

Till date several variants of process algebra have been proposed. The variant of
process algebra that we use in this paper is called *stochastic π calculus* [9,1,10].
π *calculus* can describe concurrent processes whose *configuration may change
during interaction*. Stochastic π-calculus is a type of π calculus where *stochastic
rates are imposed on the processes*.

The main constructs of stochastic π-calculus that are used in this paper are:
(1) *Parallel Composition*, (also known as *concurrency*), defined as two processes
P and Q executing concurrently and denoted as $P|Q$, (2) *Process interactions*
(also known as *communication*), defined as two processes interacting through
channels, shared by the interacting processes. The possible process interactions
π are (i) *delays* at a rate r, (ii) *input* on channel a at a rate r (denoted as $?a_{(r)}$)
and (iii) *output* on channel a at a rate r (denoted as $!a_{(r)}$). *Stochastic delay
operation* is the event of reordering a process at a rate r (denoted as $\tau_{(r)}$) while
in *complementary synchronous interactions* processes can interact by performing
complementary input and output on a common channel at a rate r (denoted by
$a_{(r)}$).

In this paper, we use the *interacting stochastic automata* version of stochastic π calculus. In other words, in this model of computation, each automaton

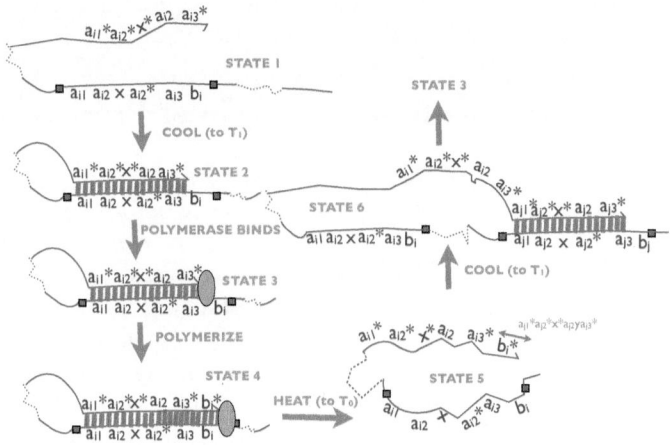

Fig. 2. Stochastic Delay Operation with π-WPCR machines: (State 1) Original π-WPCR strand just after completion of the last state transition (system temperature T_0), (State 2) When cooled to T_1 below the melting temperature of the $3'$ end of the machine but above the melting temperature of all the local stem loops in the rules and the stem-loop at the $3'$ end, the former binds to the current state of rule R_i , (State 3) Polymerase binds to the $3'$ end of the machine, (State 4) Polymerase extends the $3'$ end of the machine to copy the next state in R_i (b_i), (State 5) Strand is denatured by the application of heat (system temperature raised to T_0), (State 6) The $3'$ end of the machine now encodes $b_i^* = a_{j1}^* a_{j2}^* x^* a_{j2} a_{j3}^*$ and it can bind to the current state of R_j which is encoded as $a_{j1} a_{j2} x a_{j2}^* a_{j3}$

represents a process. These automata can be either executing a stochastic delay transition or multiple automata can be interacting. For instance, an automaton in state A can move to state A' at a specified rate r by a spontaneous delay transition $\tau @ r$. There can also be two other kinds of automata: the ones in state A can perform an input $?a$ on a channel a, and move to state A', provided that each can coordinate its transition with another automaton in state B that at the same time performs an output $!a$ on the same channel a, to move to state B'. Each channel a has an associated rate r represented as $a @ r$.

3   Original Whiplash PCR System

Before we can describe how we utilize the design principles of a WPCR machine to design a π-WPCR machine we need to first describe the original machine. In the original WPCR machine, the transition table is encoded on a single stranded DNA, W as $S - a_1 - b_1 - S - a_2 - b_2 - \ldots - S - a_n - b_n$ where each pair $a_i - b_i$ represents the transition from state a_i to state b_i. Here and in the rest of the paper, any symbol s encodes for a DNA sequence and s^* encodes for its complementary sequence. The stopper sequence S isolates one state transition rule from another. The $3'$ end of the same strand encodes the current state. For the description of the rest of the protocol, refer to Figure 1.

4   A WPCR Machine Representation of Process Algebra

The π-WPCR machine is encoded as $a_{11}a_{12}xa_{12}^{*}a_{13}b_1S\ldots a_{n1}a_{n2}xa_{n2}^{*}a_{n3}b_nS$ (in a single strand) where each rule R_i is encoded as $a_{i1}a_{i2}xa_{i2}^{*}a_{i3}b_i$ (with $a_{i1}a_{i2}xa_{i2}^{*}a_{i3}$ as the current state and b_i as the next state) and the current state of the machine encoded at the $3'$ end of the strand as $a_{i1}^{*}a_{i2}^{*}x^{*}a_{i2}a_{i3}^{*}$ (let us assume that the current state of the machine corresponds to the current state of rule R_i). In the current state encoded as $a_{i1}a_{i2}xa_{i2}^{*}a_{i3}$ in each rule R_i, the most important segment is the encoding of a stem-loop $a_{i2}xa_{i2}^{*}$. The encoding a_{i3} is used for information exchange in case of process interaction. Furthermore, the next state b_i is also the current state of a different rule R_j and is represented as $b_i = a_{j1}a_{j2}xa_{j2}^{*}a_{j3}$. We use a shorthand notation for clarity in the figures. A compact representation of a rule is shown in Figure 3(Right).

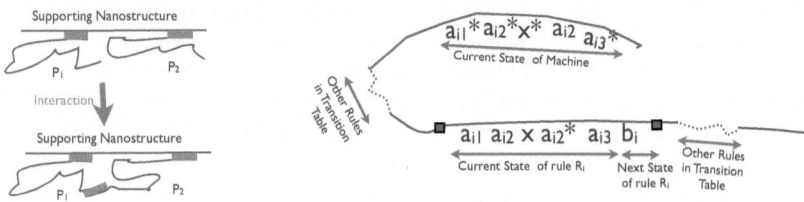

Fig. 3. (Left)Supporting nanostructure holding two processes, represented by single stranded DNA, in close proximity (via hybridization) that facilitates interprocess communication, (Right) Shorthand representation of the modified WPCR strand simply showing its current state and the rule with which it is going to bind with next. The current state of the machine and rule R_i is indicated (with current state $a_{i1}a_{i2}xa_{i2}^{*}a_{i3}$ and next state b_i). The $a_{i2}xa_{i2}^{*}$ encoding in the current state of each R_i can form a stem-loop. A similar stem-loop encoding is hidden in b_i as well. It is not shown for clarity of representation. It is this loop that contributes to the loop at the $3'$ end of the machine after a state transition. The stem-loop near the $3'$ end of the machine is essential for process communication. The other rules are not shown for clarity and their presence is indicated by the dotted portion of the strand. Every adjacent pair of rules on the transition table are separated by a stopper sequence indicated by a dark square.

5   Implementation of Stochastic Delay Operation

A *stochastic delay operation* in π-WPCR machine is the same as the state transition event in the original WPCR machine and is simulated in exactly the same manner. However, its success rate is crucially dependent on the multi-temperature thermal cycle. Refer to Figure 2 for pictorial reference on the complete process. In spite of the stem-loop encodings $a_{i2}xa_{i2}^{*}$ in each rule R_i and

$a_{i2}^* x^* a_{i2}$ at the $3'$ end of the strand, the loops are not preferentially formed when the system is cooled to T_1 above their melting temperature. Instead, the $3'$ end (encoded as $a_{i1}^* a_{i2}^* x^* a_{i2} a_{i3}^*$) of the machine binds with the current state of R_i ($a_{i1} a_{i2} x (a_{i2})^* a_{i3}$). As usual, the polymerase can attach to the $3'$ end of the machine and copy the next state from R_i. Next the strand is denatured by applying heat (meaning that the temperature of the solution is raised to T_0). The $3'$ end of the π-WPCR strand has b_i^* encoded in it. b_i^* is equivalent to $a_{j1}^* a_{j2}^* x^* a_{j2} a_{j3}^*$ which can now bind to the current state of rule R_j which is encoded as $a_{j1} a_{j2} x (a_{j2})^* a_{j3}$ and the chain of events repeats to execute another state transition. Such a state transition is referred to as a *stochastic delay operation* in interacting stochastic automata.

6   Implementation of Parallel Composition of Processes

Parallel composition is the simplest of all operators to implement with π-WPCR machines. π-WPCR machines run in isolation by virtue of its design. Hence, implementation of parallel composition with π-WPCR machines, do not need any biochemical techniques other than the ones already being used. We have already mentioned that in WPCR machines both program and its input are internal to the machine, thus allowing parallel execution of distinct programs. Thus, it is possible for two processes P_1 and P_2 to have the same transition table but different current states and yet run in parallel without interfering with the smooth execution of the other process.

7   Implementation of Process Communication

In this paper, the type of *process interaction or communication* we implement is *complementary synchronous interaction*. Suppose we have two π-WPCR machines that are executing state transitions independent of one another, until at one point one machine, say M_1 (corresponding to process P_1) cannot bind to any of the current state of the rules in its transition table. It can, however, resume its normal stochastic delay operation after it copies at the $3'$ end an encoding that matches the current state of one of its rewrite rules from another machine (corresponding to process P_2) M_2's $3'$ end. Additionally, for complementary synchronous interaction, we have to implement M_2 as a machine where it has to provide the information that M_1 needed before it can proceed with its own state transitions. In case of asynchronous interaction M_2 could have proceeded on its own or choose to wait until its current state encoded at its $3'$ end is modified while communicating with M_1. Consequently, M_2 needs to wait for M_1 to finish receiving the desired information. As mentioned earlier, by the rules of communication in process algebra, M_2 not only modifies M_1 but is modified in the process as well. Hence, we implement M_2 waiting for M_1, by encoding in M_1's $3'$ end, an encoding that matches the current state of a rule in M_2 and that is missing from the next state of all the rules in M_2. Thus, after a certain number of state transitions (need not be the same number as M_1's), M_2's encoding in

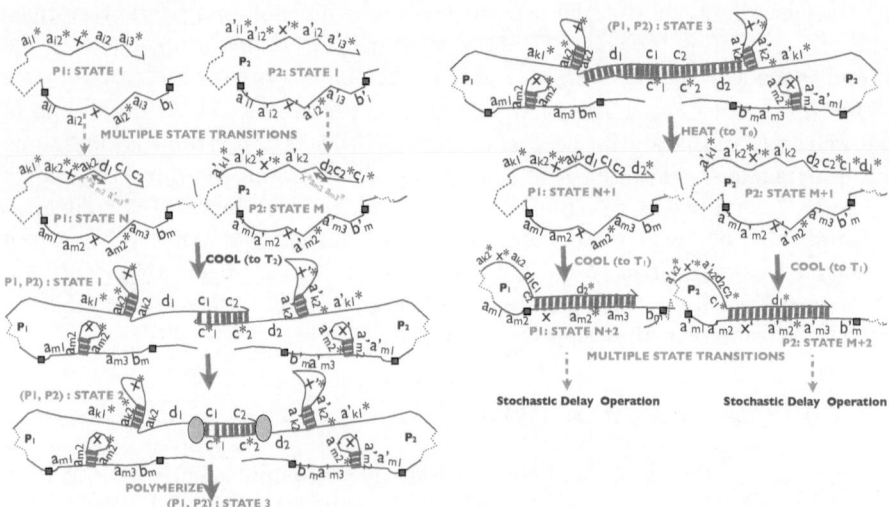

Fig. 4. π-WPCR implementation of interprocess communication: (P1: State 1) Initial state of process P_1, (P2: State 1) Initial state of process P_2, (P1:State N) State of P_1 after N transitions. The $3'$ end of P_1 encodes $d_1c_1c_2$. d_1 encodes for part of the current state in $(R_m)'$ in P_2 since it is represented as $xa^*_{m2}a_{m3}$. The dashed arrow between States P1:State 1 and P1:State N is used to denote multiple state transitions, (P2:State M) State of P_2 after M transitions. Its $3'$ end encoded as $d_2c_2^*c_1^*$ Now d_2 encodes for part of the current state in R_m in P_1 represented as $xa^*_{m2}a_{m3}$, (P1,P2: State 1) The $3'$ ends of P_1 and P_2 hybridize when cooled to T_2 below the melting temperature of not only c_1c_2 but also the local stem loops. Consequently, stem loops are formed near the $3'$ ends of the machines as well as the rules. The lower temperature is indicated by representing this particular "COOL" operation in blue, (P1,P2: State 2) A polymerase attaches to the $3'$ end in each of the machines, (P1,P2: State 3) In presence of the polymerase, the $3'$ end of P_1 copies d_2 from P_2 and the $3'$ end of P_2 copies d_1 from P_1, (P1: State N+1) Once heated to T_0, the interacting processes can separate and P_1's $3'$ end d_2^* now partly encodes for the current state of R_m, (P2: State M+1) Similar to P_1, after dehybridizing from the P_1P_2 complex, P_2's $3'$ end encoded as d_1^* can partly bind to the current state of R_m', (P1:State N+2) P_1's $3'$ end binds to R_m as a first step to a stochastic delay operation, (P2:State M+2) P_2's $3'$ end binds to R_m' as a first step to a stochastic delay operation.

its $3'$ end does not correspond to any of the current states in its rules. On the contrary, it is the same encoding that M_1 has at its $3'$ end when it needs to copy the current state encoding from M_2's $3'$ end. Consequently, the machines cannot proceed without interacting. Refer to Figure 4 for the complete protocol.

8 Discussion

The implementation of π-WPCR machine described in this paper is based on the original WPCR machine and, hence, it particularly suffers from all its drawbacks.

Nevertheless, the implementation is quite powerful in the sense that it is easy to extend the design for two processes to that for multiple processes. For instance, extending parallel composition to multiple processes does not involve any modification in the encoding of the π-WPCR machines. However, all interacting processes have to be physically close in order to communicate. Consequently, they should share a supporting nanostructure.

In summary, in this paper we presented a design for biochemical implementations of the key primitives in process algebra using modified WPCR machines. The main constructs in this model of computation are *sending data along a channel, receiving data along a channel, creating channels* and *running processes in parallel*. Since *input* and *program* are *local* to a WPCR machine, multiple processes can be executed *concurrently* in the original WPCR machines. Our implementation of interprocess communication using π-*WPCR machines* simulates rest of the constructs mentioned in Section 2. One challenging open question is whether we can use isothermal protocols [11] since the former has evidently more flexibility of operations. The biggest hindrance to using an isothermal WPCR machine for implementing process algebra is that the isothermal protocol uses a strand displacing polymerase and this is not suitable for our implementation of process communication.

References

1. Milner, R.: Communicating and mobile systems: the pi calculus. Cambridge Univ. Press, Cambridge (1999)
2. Cardelli, L.: On process rate semantics. Theo. Comp. Sci. 391(3), 190–215 (2008)
3. Bassler, B.: How bacteria talk to each other: regulation of gene expression by quorum sensing. Curr. Opin. Microbiol. 2, 582–587 (1999)
4. Koeneman, A., Capco, D.G.: A Kaleidoscope of Modern Life Sciences and Modern Medicine. Encyclopedia of Molecular Cell Biology and Molecular Medicine, vol. 2. Ramtech Ltd., Larkspur (2005)
5. Winfree, E.: Simulation of computing by self-assembly. Technical Report 1998.22, Caltech (1998)
6. Sakamoto, K., Kiga, D., Komiya, K., Gouzu, H., Yokoyama, S., Ikeda, S., Sugiyama, H., Hagiya, M.: State transitions by molecules. Biosystems 52(1), 81–91 (1999)
7. Dirks, R.M., Pierce, N.A.: Triggered Amplification of Hybridization Chain Reaction. PNAS 101(43), 15275–15278 (2004)
8. Berg, J., Tymoczko, J., Stryer, L.: Molecular Cell Biology. W.H. Freeman, New York (2002)
9. Phillips, A., Cardelli, L.: A correct abstract machine for the stochastic pi-calculus. In: Bioconcur. (2004)
10. Phillips, A., Yoshida, N., Eisenbach, S.: A distributed abstract machine for boxed ambient calculi. In: Schmidt, D. (ed.) ESOP 2004. LNCS, vol. 2986, pp. 155–170. Springer, Heidelberg (2004)
11. Reif, J.H., Majumder, U.: Computing with Isothermal Autocatalytic Whiplash PCR. In: DNA 14. LNCS. Springer, Heidelberg (2008)

NP-Completeness of the Direct Energy Barrier Problem without Pseudoknots

Ján Maňuch[1], Chris Thachuk[2], Ladislav Stacho[1], and Anne Condon[2]

[1] School of Computing Science and Department of Mathematics,
Simon Fraser University, Canada
jmanuch@sfu.ca

[2] Department of Computer Science, University of British Columbia, Canada

Abstract. Knowledge of energy barriers between pairs of secondary structures for a given DNA or RNA molecule is useful, both in understanding RNA function in biological settings and in design of programmed molecular systems. Current heuristics are not guaranteed to find the exact energy barrier, raising the question whether the energy barrier can be calculated efficiently. In this paper, we study the computational complexity of a simple formulation of the energy barrier problem, in which each base pair contributes an energy of -1 and only base pairs in the initial and final structures may be used on a folding pathway from initial to final structure. We show that this problem is NP-complete.

1 Introduction

We study the computational complexity of the energy barrier problem for nucleic acids: what energy barrier must be overcome for a DNA or RNA molecule to adopt a given final secondary structure, starting from a given initial secondary structure? We first provide some motivation for studying the energy barrier problem, then describe a simple formulation of the problem and summarize our results.

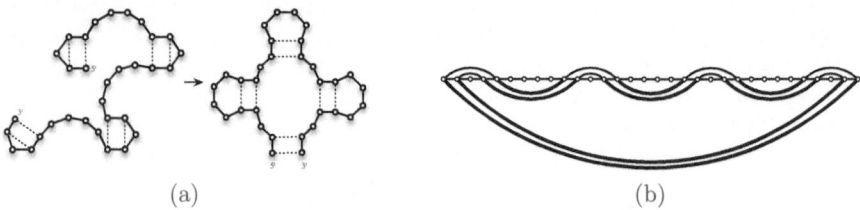

(a)   (b)

Fig. 1. (a) An initial secondary structure (left) and a final secondary structure (right) for a given RNA strand. (b) A corresponding arc diagram. The top of the diagram denotes the base pairs in the initial structure while the bottom of the diagram, shown with bold arcs, denotes the base pairs of the final structure.

Motivation. Methods for calculating energy barriers are useful, in both rational design of programmed nucleic acid systems and in understanding the mechanisms

R. Deaton and A. Suyama (Eds.): DNA 15, LNCS 5877, pp. 106–115, 2009.

of RNA function in the cell. This is because, both in the design and biological
contexts, secondary structure and folding pathways are central to function. Many
designed nucleic acid systems rely critically on the premise that the constituent
molecules will follow certain folding pathways and avoid others [1,2,3,4,5,6,7].
Designs of such systems typically ensure that the desired folding pathway has
a low energy barrier, compared with alternatives. While this property can be
straightforward to establish for simple molecular systems, a method for energy
barrier calculation would be useful when verifying that a system of large or even
moderate scale has the desired behaviour [8]. In the biological context, knowledge
of energy barriers between intermediate structures on the pathway from the open
chain to the folded configuration of biological molecules is useful in determining
folding efficiency and structure [9,10,11,12].

Methods for simulation of DNA or RNA folding pathways often estimate
energy barriers between secondary structures, in order to calculate probabilities
of transitioning between structures which are included in maps of the energy
folding landscape [13,14,15,16]. Heuristics for energy barrier calculation are also
used to construct barrier trees, which are helpful in visualizing energy landscapes
[17], and to elucidate properties of disordered systems in statistical physics [18].

In light of its importance, it's natural to ask: what is the computational complexity of the energy barrier problem for nucleic acid secondary structures? An
efficient algorithm for the problem might be a valuable replacement of currently-
used heuristics in the applications mentioned above. On the other hand, intractability of the barrier problem would suggest that the answer to another
complex problem might be determined from observations of nucleic acid folding
pathways of a sequence constructed for the complex problem.

The Energy Barrier Problem. We formulate the problem as follows. A *secondary structure* T for an RNA molecule of length n is a set of base pairs $i.j$,
with $1 \leq i < j \leq n$, such that (i) each base index i or j appears in at most
one base pair and (ii) the bases at indices i and j form a Watson-Crick (i.e.,
C-G, A-U, or A-T) base pair. Since we represent secondary structures using arc
diagrams, we use the word *arc* interchangeably with base pair (see Fig. 1). Our
main results pertain to pseudoknot free secondary structures, that is, structures
with no crossing arcs — see Section 2 for definitions. We assume a very simple
energy model for secondary structures in which each arc contributes an energy
of -1. Thus, as is roughly consistent with more realistic energy models, the more
base pairs in a structure the lower its energy. We denote the energy of secondary
structure T by $E(T)$.

A *folding pathway* is a sequence of pseudoknot free secondary structures for
a given molecule, each of which differs from its predecessor by the addition or
removal of one arc . The *energy barrier* of a folding pathway $T_0, T_1, \ldots, T_r$ is the
maximum of $E(T_i) - E(T_0)$, where the max is taken over i in the range $1 \leq i \leq r$.
Note that there is always a folding pathway from $\mathcal{I}$ to $\mathcal{F}$, in which first all arcs
of $\mathcal{I}$ are removed and then all arcs of $\mathcal{F}$ are added. All secondary structures on
such a pathway are pseudoknot free since they are either subsets of $\mathcal{I}$ or of $\mathcal{F}$,
both of which are pseudoknot free. However, the energy barrier of this pathway

is $|\mathcal{I}|$. The question is whether there is another pathway that avoids such a high energy barrier, by adding arcs of $\mathcal{F}$ before all arcs of $\mathcal{I}$ are removed. The *energy barrier problem* is to determine whether there is a folding pathway from a given initial structure $\mathcal{I}$ to a given final structure $\mathcal{F}$, whose energy barrier is at most k, for some given k.

The results presented here are a first step towards solving the energy barrier problem. Our results pertain to restricted types of folding pathways, namely *direct* folding pathways. Such pathways were introduced by Morgan and Higgs [18]. A folding pathway from secondary structure $\mathcal{I}$ to $\mathcal{F}$ is *direct* if the only arcs which are added are those from $\mathcal{F} - \mathcal{I}$ and the only arcs which are removed are those from $\mathcal{I} - \mathcal{F}$. Thus, there are exactly $|\mathcal{I} - \mathcal{F}| + |\mathcal{F} - \mathcal{I}|$ steps along a direct folding pathway. All of the designed nucleic acid folding pathway systems with which we are familiar are such that the desired folding pathway is direct [2,3,4,7]. The *energy barrier problem for direct pseudoknot free folding pathways* (DPKF-EB PROBLEM) is: given initial and final pseudoknot free secondary structures $\mathcal{I}$ and $\mathcal{F}$ and an integer k, is there a direct folding pathway from $\mathcal{I}$ to $\mathcal{F}$ which has energy barrier at most k? In Section 3, we show that the DPKF-EB problem is NP-complete.

The rest of the paper is organized as follows. We provide definitions of pathways, energy barrier, and other useful notation in Section 2. We prove our result in Section 3: Theorem 2 shows that the energy barrier problem for direct folding pathways of pseudoknot free structures is NP-complete. We conclude with a brief discussion of our results.

2   Definitions

Fix initial and final pseudoknot free secondary structures $\mathcal{I}$ and $\mathcal{F}$. A *direct pseudoknot free folding pathway* from $\mathcal{I}$ to $\mathcal{F}$ is a sequence of pseudoknot free secondary structures $\mathcal{I} = \mathcal{T}_0, \mathcal{T}_1, \ldots, \mathcal{T}_r = \mathcal{F}$, where each $\mathcal{T}_i$ is obtained from $\mathcal{T}_{i-1}$ by either the addition of one arc from $\mathcal{F} - \mathcal{I}$ or the removal of one arc from $\mathcal{I} - \mathcal{F}$. We call each such addition or removal an *arc operation* and we let $+x$ and $-x$ denote the addition and removal of the arc x, respectively. The $\mathcal{T}_i$'s are called *intermediate structures*. A folding pathway can thus be specified by its corresponding sequence of arc operations; we call this a *transformation sequence*. A *direct pseudoknot-free transformation sequence* specifies a folding pathway which is both direct and pseudoknot free.

The *energy barrier* of a folding pathway $\mathcal{I} = \mathcal{T}_0, \mathcal{T}_1, \ldots, \mathcal{T}_r = \mathcal{F}$ is the maximum of $E(\mathcal{T}_i) - E(\mathcal{I})$, where the max is taken over all integers i in the range $1 \leq i \leq r$. The *energy difference* of each intermediate configuration $\mathcal{T}_i$ is defined as $E(\mathcal{T}_i) - E(\mathcal{I})$. If Π is the transformation sequence for this pathway, then the *energy barrier* of transformation sequence Π, denoted as $\Delta E(\mathcal{I}, \mathcal{F}, \Pi)$, is defined to be the energy barrier of the corresponding folding pathway.

In our result, it is convenient to work with weighted arcs. To motivate why, note that the union $\mathcal{I} \cup \mathcal{F}$ of two pseudoknot free secondary structures may be pseudoknotted, i.e., may have crossing arcs, even when both $\mathcal{I}$ and $\mathcal{F}$ are pseudoknot free. In a pseudoknotted structure, we use the term *band* to refer to a set of nested arcs, each of which crosses the same set of arcs. In a folding pathway from $\mathcal{I}$ to $\mathcal{F}$ which minimizes the energy barrier, we can assume without loss of generality that when one arc in a band of $\mathcal{I} \cup \mathcal{F}$ is added, then all arcs in the band are added consecutively. Similarly, we can assume without loss of generality that when one arc in a band is removed, then all arcs in the band are removed consecutively. Thus, it is natural to represent the set of arcs in a band as one arc with a weight equal to the number of arcs in the band.

Hence we generalize the notion of secondary structure as follows. A *weighted arc* $I = (I^b, I^e)^{I^w}$ is specified by start and end indices $I^b < I^e$ and a weight I^w. We say that two weighted arcs I and J are *crossing* if either $I^b \leq J^b$ and $I^e \leq J^e$, or $J^b \leq I^b$ and $J^e \leq I^e$. A *configuration* is a set of weighted arcs. Configuration $\{I_i\}_{i=1}^n$ is *pseudoknot-free* if for all $1 \leq i < j \leq n$, I_i and I_j are not crossing. The energy of configuration $\mathcal{I} = \{I_i\}_{i=1}^n$ is $\mathrm{E}(\mathcal{I}) = -\sum_{i=1}^n I^w$. The previous definitions can easily be generalized to weighted arcs.

Finally, we define the main problem studied in this paper, namely the DPKF-EB problem, and also the 3-PARTITION problem which is used to show NP-completeness of DPKF-EB.

DPKF-EB PROBLEM (Energy barrier problem for direct folding pathways without pseudoknots). Given two pseudoknot-free configurations $\mathcal{I} = \{I_i\}_{i=1}^n$ (initial) and $\mathcal{F} = \{F_i\}_{i=1}^m$ (final), and integer k, is there a direct pseudoknot-free transformation sequence S such that the energy barrier of S is at most k, i.e., $\Delta\mathrm{E}(\mathcal{I}, \mathcal{F}, S) \leq k$.

3-PARTITION PROBLEM. Given $3n$ integers $a_1, \ldots, a_{3n}$ such that $a_1 + \cdots + a_{3n} = nA$ and $A/4 < a_i < A/2$ for each i, the 3-PARTITION PROBLEM asks: is there a partition of the integers $\{1, \ldots, 3n\}$ into disjoint triples $\mathrm{G}_1, \mathrm{G}_2, \ldots, \mathrm{G}_n$ such that the sum of a_j, where j belongs to G_i is equal to A, i.e., $\mathrm{c}(\mathrm{G}_i) = \sum_{j \in \mathrm{G}_i} a_j = A$ for each $i = 1, \ldots, n$.

Theorem 1 (Garey, Johnson (1979) [19]). *The* 3-PARTITION *problem is NP-complete even if A is polynomial in n.*

Note that the 3-PARTITION problem is in P if A is a constant.

3   Result

Theorem 2. *The* DPKF-EB   *problem, namely the energy barrier problem for direct folding pathways without pseudoknots, is NP-complete.*

Proof. It is straightforward to show that the DPKF-EB problem is in NP. Given an instance $(\mathcal{I}, \mathcal{F}, k)$, it is sufficient to non-deterministically guess a direct folding pathway from $\mathcal{I}$ to $\mathcal{F}$, and to verify that the energy barrier of this path is at most k. Note that the length of any such pathway is at most $|\mathcal{I}| + |\mathcal{F}|$.

To show that the DPKF-EB problem is NP-hard, we provide a reduction from the 3-PARTITION problem. We first provide a formal description of the reduction, and then prove correctness in detail.

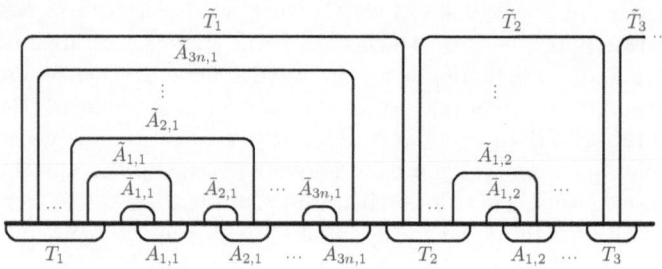

Fig. 2. Organization of weighted arcs in the initial (top) and the final (bottom) configurations

Consider an instance of the 3-PARTITION problem $A/2 > a_1 \geq \cdots \geq a_{3n} > A/4$ such that $\sum_{j=1}^{3n} a_j = nA$ and A is polynomial in n. We define an instance $(\mathcal{I}, \mathcal{F}, k)$ of the DPKF-EB problem as follows. The initial configuration $\mathcal{I}$ contains weighted arcs $\{\bar{A}_{j,i}; \ j = 1, \ldots, 3n, \ i = 1, \ldots, n\} \cup \{\tilde{A}_{j,i}; \ j = 1, \ldots, 3n, \ i = 1, \ldots, n\} \cup \{\tilde{T}_i; \ i = 1, \ldots, n\}$. The final configuration $\mathcal{F}$ is $\{A_{j,i}; \ j = 1, \ldots, 3n, \ i = 1, \ldots, n\} \cup \{T_i; \ i = 1, \ldots, n\}$. The arcs are organized as in Figure 2.

Formally, the arcs are organized as follows:

$$T_1^b < \tilde{T}_1^b < \tilde{A}_{3n,1}^b < \cdots < \tilde{A}_{1,1}^b < T_1^e < \bar{A}_{1,1}^b,$$
$$T_i^b < \tilde{T}_{i-1}^e < \tilde{T}_i^b < \tilde{A}_{3n,i}^b < \cdots < \tilde{A}_{1,i}^b < T_i^e < \bar{A}_{1,i}^b, \ \forall i = 2, \ldots, n,$$
$$\bar{A}_{j,i}^b < A_{j,i}^b < \bar{A}_{j,i}^e < \tilde{A}_{j,i}^e < A_{j,i}^e, \ \forall i = 1, \ldots, n, \ j = 1, \ldots, 3n,$$
$$A_{j,i}^e < \bar{A}_{j+1,i}^b, \ \forall i = 1, \ldots, n, \ j = 1, \ldots, 3n - 1,$$
$$A_{3n,i}^e < T_{i+1}^b, \ \forall i = 1, \ldots, n - 1,$$
$$A_{3n,n}^e < \tilde{T}_n^e.$$

The weights of arcs are set up as follows. For all $i = 1, \ldots, n$ and $j = 1, \ldots, 3n$: $\tilde{A}_{j,i}^w = 4ia_j, \bar{A}_{j,i}^w = k - (j-1)A - 4ia_j, A_{j,i}^w = k - jA$. Also,

$$\tilde{T}_1^w = k - (7n - 4)A,$$
$$\tilde{T}_i^w = k - (6n + 8)nA - 4(n-1)iA, \ \forall i = 2, \ldots, n,$$
$$T_i^w = k - (6n + 8)nA, \ \forall i = 1, \ldots, n - 1,$$
$$T_n^w = k,$$

where $k > 4(5n^2 + n + 1)A$ is the energy barrier.

We now prove that the DPKF-EB instance has a solution with energy barrier at most k if and only if the 3-PARTITION instance $a_1, \ldots, a_{3n}$ has a solution.

First, assume that the 3-PARTITION instance has a solution $G_1, \ldots, G_n$, where $G_i = \{j_{i,1}, j_{i,2}, j_{i,3}\}$. Let $f(j) = i$ if $j \in G_i$, for every $j = 1, \ldots, 3n$. We will show that the transformation sequence

$$-\bar{A}_{1,f(1)}, -\tilde{A}_{1,f(1)}, +A_{1,f(1)}, \ldots, -\bar{A}_{3n,f(3n)}, -\tilde{A}_{3n,f(3n)}, +A_{3n,f(3n)}, \quad (1)$$

$$\underbrace{-\tilde{A}_{1,1}, \ldots, -\tilde{A}_{3n,1}}_{\text{without } -\tilde{A}_{j_1,1,1}, -\tilde{A}_{j_1,2,1}, -\tilde{A}_{j_1,3,1}} \quad , -\tilde{T}_1, +T_1, \ldots,$$

$$\underbrace{-\tilde{A}_{1,n}, \ldots, -\tilde{A}_{3n,n}}_{\text{without } -\tilde{A}_{j_n,1,n}, -\tilde{A}_{j_n,2,n}, -\tilde{A}_{j_n,3,n}} \quad , -\tilde{T}_n, +T_n, \quad (2)$$

$$\underbrace{-\bar{A}_{1,1}, +A_{1,1}, -\bar{A}_{1,2}, +A_{1,2}, \ldots, -\bar{A}_{3n,n}, +A_{3n,n}}_{\text{without indexes } 1, f(1); \ 2, f(2); \ \ldots; \ 3n, f(3n)} \quad (3)$$

is pseudoknot-free with energy barrier exactly k. For clarity, the $-$ sign marks the arcs from the initial configuration which are being removed and the $+$ sign marks the arcs from the final configuration which are being added. It is easy to see that the sequence is pseudoknot-free, since

- each $A_{j,i}$ is crossing only $\tilde{A}_{j,i}$ and $\bar{A}_{j,i}$ and is added only when these two arcs are already removed; and
- each T_i is crossing the following arcs in the initial configuration: $\tilde{T}_{i-1}$ (if $i > 2$), $\tilde{T}_i$ and $\tilde{A}_{1,i}, \ldots, \tilde{A}_{3n,i}$ and they are all removed before T_i is added.

Second, let us verify that the energy difference of each intermediate configuration is at most k. First, in line (1), by induction, for each $j = 1, \ldots, 3n$, before removing $-\bar{A}_{j,f(j)}, -\tilde{A}_{j,f(j)}$ the energy difference is $(j-1)A$ and after removal it is k. Then after adding $+A_{j,f(j)}$ it decreases to jA. At the end of line (1), the energy difference is $3nA$. Next, we need to check that the sum of weights of arcs

$$\underbrace{-\tilde{A}_{1,1}, \ldots, -\tilde{A}_{3n,1}}_{\text{without } -\tilde{A}_{j_1,1,1}, -\tilde{A}_{j_1,2,1}, -\tilde{A}_{j_1,3,1}} \quad , -\tilde{T}_1$$

is at most $k - 3nA$. The sum of weights of these arcs is exactly

$$\sum_{j=1}^{3n} \tilde{A}_{j,1}^w - \sum_{\ell=1}^{3} \tilde{A}_{j_1,\ell,1}^w + \tilde{T}_1^w = \sum_{j=1}^{3n} 4a_j - \sum_{\ell=1}^{3} 4a_{j_1,\ell} + k - 7nA + 4A$$

$$= 4nA - 4A + k - 7nA + 4A = k - 3nA.$$

Thus, just before adding $+T_1$, the energy difference is again exactly k. And after adding $+T_1$, it is $6n^2A+8nA$. Similar calculations show that the energy difference will alternate between k (after each removal subsequence) and $6n^2A+8nA$ (after each addition of $+T_i$) in line (2) with exception of the last addition, when the energy difference is 0. In line (3), all remaining arcs from the initial configuration $(-\bar{A}_{j,i})$ are removed and all remaining arcs from the final configuration $(+A_{j,i})$ are added. Note that each removal is possible since $\bar{A}_{j,i}^w < k$ and after processing each pair $-\bar{A}_{j,i}, +A_{j,i}$, energy difference only decreases since $\bar{A}_{j,i}^w - A_{j,i}^w = A - 4ia_j < 0$.

Now, assume that there is a pseudoknot-free transformation sequence S with the energy barrier at most k. From S, we will construct a solution for the original

3-PARTITION instance and show that it is a valid solution. We organize our proof into three parts, in line with the three properties described in the intuition at the start of the proof.

Consider the subsequence of S containing only additions, i.e., arcs from the final configuration. Let S^+ denote this subsequence. We assume without loss of generality that all removals in S happen only when needed, i.e., the next addition would not be possible without those removals. Hence, the subsequence S^+ determines the whole sequence S. By *processing* an arc $+I$ in S^+ we mean removal of all arcs $-J$ in S immediately preceding $+I$ (that is, not preceding any other $+I'$ appearing in S^+ before $+I$) and adding $+I$.

The first part of our proof considers the prefix of S^+ just before the first T_ℓ is added. Let this prefix be:

$$+A_{j_1,i_1}, +A_{j_2,i_2}, \ldots, +A_{j_M,i_M} \tag{4}$$

where M is the number of $+A_{j,i}$'s added before $+T_\ell$. We use this prefix to define a potential solution to the 3-PARTITION problem:

$$G_i = \{j_\ell;\ i_\ell = i\},$$

for every $i = 1, \ldots, n$.

Ultimately we will show that the G_i's (or a slight perturbation of the G_i's) form a solution to the 3-PARTITION problem. Towards this goal, our first two lemmas below prove some useful properties of the G_i's that can be inferred from the weights of the arcs in the folding pathway prefix (4) and the corresponding removed arcs. Let $|G_i|_j$ denote the number of elements in G_i with value at most j. In order for the G_i's to be a valid solution, $|G_i|_j$ should be exactly j for all $j, 1 \le j \le 3n$. Moreover it should be the case that $\sum_{i=1}^{n} c(G_i) = nA$ where $c(G_i)$ denotes the sum of a_j for $j \in G_i$ (see the definition of 3-PARTITION). The statements of the two lemmas below assert somewhat weaker properties of the G_i's.

Lemma 1. *For every* $j = 1, \ldots, 3n$, $\sum_{i=1}^{n} |G_i|_j \le j$. *Consequently,* $M \le 3n$.

Proof. Let $+T_\ell$ be the first $+T_i$ in S^+. Consider an $+A_{j,i}$ appearing before $+T_\ell$. Recall that before adding $+A_{j,i}$, we need to remove both $-\tilde{A}_{j,i}$ and $-\bar{A}_{j,i}$. Since, $\tilde{A}_{j,i}^w + \bar{A}_{j,i}^w = k - (j-1)A$, the energy difference has to be at most $(j-1)A$ for $+A_{j,i}$ to be added. Note that processing of each $+A_{j,i}$ appearing in S^+ before $+T_\ell$ will increase the energy difference by A, as it requires both $-\tilde{A}_{j,i}$ and $-\bar{A}_{j,i}$ to be removed first and $\bar{A}_{j,i}^w + \tilde{A}_{j,i}^w - A_{j,i}^w = k - (j-1)A - 4ia_j + 4ia_j - (k - jA) = A$. For instance, an $+A_{1,i}$ can only appear at the first position of the part of the subsequence S^+ before $+T_\ell$, since it requires the energy difference at least 0 and after any $+A_{j,i}$ is added, the energy difference increases to A. Thus, starting from the second position, no $+A_{1,i'}$ can be added before $+T_\ell$. Similarly, $+A_{j,i}$ can appear only in the first j positions of the subsequence of S^+ before $+T_\ell$. The lemma easily follows.

In the next lemma we we use double brackets to denote multisets: for example $\{\{1,2,2\}\}$ is the multiset with elements 1, 2, and 2 and $\{\{1,1,2\}\} \neq \{\{1,2,2\}\}$.

Lemma 2. $\sum_{i=1}^{n} c(G_i) \leq nA - (3n - M)A/4$, where the equality happens only if $M = 3n$ and $\{\{a_{j_1}, \ldots, a_{j_M}\}\} = \{\{a_1, \ldots, a_{3n}\}\}$.

Proof. The proof will be included in the full version of the paper.

We now turn to the second part of our proof: we show that, following the initially-added sequence of $+A_{i,j}$'s, the T_i's must be added in increasing order of i. That is, the arcs $+T_1, \ldots, +T_n$ appear in the subsequence S^+ consecutively (with no $+A_{j,i}$ in between) and in this order. The next lemma shows that the first $+T_i$ in the sequence S^+ must be $+T_1$ and the following lemma reasons about the rest of the sequence of $+T_i$'s.

Lemma 3. The first $+T_i$ in S^+ is $+T_1$.

Proof. The proof will be included in the full version of the paper.

Hence, by the above lemma, the subsequence S^+ has the following form

$$+A_{j_1,i_1}, +A_{j_2,i_2}, \ldots, +A_{j_M,i_M}, +T_1$$

followed by the all remaining $+A_{j,i}$'s and $+T_i$'s. The following lemma gives more detailed insight into order of arcs in S^+.

In the remaining lemmas we adopt notation which was introduced by Graham, Knuth and Patashnik [20]:

$$[i > j] = \begin{cases} 1, & \text{if } i > j; \\ 0, & \text{otherwise,} \end{cases} \quad \text{and} \quad [i = j] = \begin{cases} 1, & \text{if } i = j; \\ 0, & \text{otherwise.} \end{cases}$$

Lemma 4. All T_i's appear in S^+ in one sequence and in increasing order.

Proof. The proof will be included in the full version of the paper.

Hence, by the above lemmas, the subsequence S^+ has the following form

$$+A_{j_1,i_1}, +A_{j_2,i_2}, \ldots, +A_{j_M,i_M}, +T_1, +T_2, \ldots, +T_n$$

followed by the all remaining $+A_{j,i}$'s.

Moving on to the last part of the proof: we show that the G_i's defined by the initial sequence of $+A_{i,j}$'s form a valid solution (or can be perturbed slightly to form a valid solution) by arguing that only in this case can all of the T_ℓ's be added without exceeding the energy barrier. Specifically, we will show that $M = 3n$ and $\{\{a_{j_1}, \ldots, a_{j_{3n}}\}\} = \{\{a_1, \ldots, a_{3n}\}\}$. For this purpose, the next two lemmas prove lower bounds on sums of the $c(G_i)$'s.

Lemma 5. For every $\ell = 1, \ldots, n$, $\sum_{i=1}^{\ell} i(c(G_i) - A) \geq (M - 3n)A/4$.

Proof. The proof will be included in the full version of the paper.

Using the inequalities from Lemma 5, we will lower bound the sum of $c(G_i)$'s.

Lemma 6. *We have $\sum_{i=1}^{n} c(G_i) \geq nA - (3n - M)A/4$, where the equality happens only if $c(G_1) = A - (3n - M)A/4$ and $c(G_i) = A$, for every $i = 2, \ldots, n$.*

Proof. The proof will be included in the full version of the paper.

By Lemmas 2 and 6, we have $\sum_{i=1}^{n} c(G_i) = nA - (3n - M)A/4$, i.e., we have equality in both Lemma 2 and Lemma 6. Thus, by Lemma 2, we have that $M = 3n$ and $\{\{a_{j_1}, \ldots, a_{j_{3n}}\}\} = \{\{a_1, \ldots, a_{3n}\}\}$.

Although this does not imply that $G_1, \ldots, G_n$ forms a decomposition of set $\{1, 2, \ldots, 3n\}$, for instance, if $a_1 = a_2$, the multiset $\{\{j_1, \ldots, j_{3n}\}\}$ could contain zero 1's and two 2's, the sets $G_1, \ldots, G_n$ could be mapped to the decomposition of $\{1, 2, \ldots, 3n\}$ just by a sequence of replacements i's with j's assuming $a_j = a_{j+1} = \cdots = a_i$. Furthermore, by Lemma 6, we have $c(G_1) = A - (3n - M)A/4 = A$ and also $c(G_i) = A$ for all $i = 2, \ldots, n$. Hence, the sets $G_1, \ldots, G_n$ (possibly modified as described above) are the solution to the 3-PARTITION problem.

The reduction is polynomial as the sum of weights of all arcs (which is the total number of arcs in the unweighted instance) is

$$\sum_{i=1}^{n} \left(\tilde{T}_i^w + T_i^w + \sum_{j=1}^{3n} (\tilde{A}_{j,i}^w + \bar{A}_{j,i}^w + A_{j,i}^w) \right) < n \cdot 2k + 3n^2 \cdot 2k = \mathcal{O}(n^2 k) = \mathcal{O}(n^4 A),$$

and A is assumed to be polynomial in n.

4   Conclusions

We have shown that the energy barrier problem for direct folding pathways is NP-complete, via a reduction from the 3-PARTITION problem. This justifies the use of heuristics for estimating energy barriers [17,14,18,16]. An interesting open question is whether there is an algorithm that is guaranteed to return the energy barrier and which works well on practical instances of the problem (while not in the worst case).

Our proof can help shed insight on energy landscapes. Consider an instance $(\mathcal{I}, \mathcal{F}, k)$ of the DPKF-EB problem which is derived from a "yes" instance of 3-PARTITION according to our construction. There are exponentially many possible prefixes (of the type shown in (4)) which could precede the addition of T_1, all of which do not exceed the energy barrier k. Of these, it may be that only one defines a valid solution of G_i's. Thus, if pathways are followed according to a random process, it could take exponential time for the random process to find the pathway with energy barrier k.

We do not resolve the computational complexity of the general energy barrier problem, in which the pathway need not be direct. Two challenges in understanding the complexity of this problem which need to be considered are repeat arcs and temporary arcs, where, by repeat arcs, we mean arcs which are added or removed multiple times along the pathway, and by temporary arcs, we mean arcs that are not in the initial or final structures of the pathway.

References

1. Kameda, A., Yamamoto, M., Uejima, H., Hagiya, M., Sakamoto, K., Ohuchi, A.: Hairpin-based state machine and conformational addressing: Design and experiment. Natural Computing 4, 103–126 (2005)
2. Yurke, B., Turberfield, A.J., Mills, A.J.J., Simmel, F.C., Neumann, J.L.: A DNA-fuelled molecular machine made of DNA. Nature 406, 605–608 (2000)
3. Seelig, G., Soloveichik, D., Zhang, D.Y., Winfree, E.: Enzyme-free nucleic acid logic circuits. Science 314, 1585–1588 (2006)
4. Simmel, F.C., Dittmer, W.U.: DNA nanodevices. Small 1, 284–299 (2005)
5. Uejima, H., Hagiya, M.: Secondary structure design of multi-state DNA machines based on sequential structure transitions. In: Chen, J., Reif, J.H. (eds.) DNA 2003. LNCS, vol. 2943, pp. 74–85. Springer, Heidelberg (2004)
6. Hagiya, M., Yaegashi, S., Takahashi, K.: Computing with hairpins and secondary structures of DNA. In: Chen, J., Jonoska, N., Rozenberg, G. (eds.) Nanotechnology: Science and Computation. Natural Computing Series, pp. 293–308. Springer, Heidelberg (2006)
7. Yin, P., Choi, H., Calvert, C., Pierce, N.: Programming biomolecular self-assembly pathways. Nature 451, 318–322 (2008)
8. Uejima, H., Hagiya, M.: Analyzing secondary structure transition paths of DNA/RNA molecules. In: Chen, J., Reif, J.H. (eds.) DNA 2003. LNCS, vol. 2943, pp. 86–90. Springer, Heidelberg (2004)
9. Chen, S.J., Dill, K.A.: RNA folding energy landscapes. Proc. Nat. Acad. Sci. 97(2), 646–651 (2000)
10. Russell, R., Zhuang, X., Babcock, H., Millett, I., Doniach, S., Chu, S., Herschlag, D.: Exploring the folding landscape of a structured RNA. Proc. Nat. Acad. Sci. 99, 155–160 (2002)
11. Shcherbakova, I., Mitra, S., Laederach, A., Brenowitz, M.: Energy barriers, pathways, and dynamics during folding of large, multidomain RNAs. Curr. Opin. Chem. Biol., 655–666 (2008)
12. Treiber, D.K., Williamson, J.R.: Beyond kinetic traps in RNA folding. Curr. Opin. Struc. Biol. 11, 309–314 (2001)
13. Flamm, C., Fontana, W., Hofacker, I.L., Schuster, P.: RNA folding at elementary step resolution. RNA, 325–338 (2000)
14. Tang, X., Thomas, S., Tapia, L., Giedroc, D.P., Amato, N.M.: Simulating RNA folding kinetics on approximated energy landscapes. J. Mol. Biol. 381, 1055–1067 (2008)
15. van Batenburg, F.H.D., Gultyaev, A.P., Pleij, C.W.A., Ng, J., Oliehoek, J.: Pseudobase: a database with RNA pseudoknots. Nucl. Acids Res. 28(1), 201–204 (2000)
16. Wolfinger, M.T.: The energy landscape of RNA folding. Master's thesis, University Vienna (2001)
17. Flamm, C., Hofacker, I.L., Stadler, P.F., Wolfinger, M.T.: Barrier trees of degenerate landscapes. Zeitschrift für Physikalische Chemie 216, 155–174 (2002)
18. Morgan, S.R., Higgs, P.G.: Barrier heights between ground states in a model of RNA secondary structure. J. Phys. A: Math. Gen. 31, 3153–3170 (1998)
19. Garey, M.R., Johnson, D.S.: Computers and Intractability: A Guide to the Theory of NP-Completeness. W. H. Freeman & Co., New York (1979)
20. Graham, R., Knuth, D., Patashnik, O.: Concrete Mathematics: a foundation for computer science. Addison-Wesley, Reading (1989)

The Effect of Malformed Tiles
on Tile Assemblies within kTAM*

Ya Meng and Navin Kashyap

Department of Mathematics and Statistics
Queen's University, Kingston, ON, K7L 3N6, Canada
{mengya,nkashyap}@mast.queensu.ca

Abstract. Many different constructions of proofreading tile sets have been proposed in the literature to reduce the effect of deviations from ideal behaviour of the dynamics of the molecular tile self-assembly process. In this paper, we consider the effect on the tile assembly process of a different kind of non-ideality, namely, imperfections in the tiles themselves. We assume a scenario in which some small proportion of the tiles in a tile set are "malformed". We study, through simulations, the effect of such malformed tiles on the self-assembly process within the kinetic Tile Assembly Model (kTAM). Our simulation results show that some tile set constructions show greater error-resilience in the presence of malformed tiles than others. For example, the 2- and 3-way overlay compact proofreading tile sets of Reif *et al.* [4] are able to handle malformed tiles quite well. On the other hand, the snaked proodreading tile set of Chen and Goel [1] fails to form even moderately-sized tile assemblies when malformed tiles are present. We show how the Chen-Goel construction may be modified to yield snaked proofreading tile sets that show good resistance to the effect of malformed tile.

1   Introduction

Molecular self-assembly is a process of bottom-up fabrication of complex structures from simple parts. A widely used mathematical model of this process is Erik Winfree's Tile Assembly Model [8],[9], which suitably extends Wang's tiling model [7] by taking into account some of the thermodynamic aspects of molecular self-assembly. The building blocks in this model are square tiles with labels and "glues" on the edges. In the *abstract Tile Assembly Model (aTAM)*, tiles attach to each other along edges with matching labels, provided the strength of the attachment (determined by the glues) exceeds a certain threshold. Tiles attaching to each other according to these rules of attachment form large assemblies. This process can be used to carry out computation, by encoding data and computational rules in the tile edge labels and glues.

The *kinetic Tile Assembly Model (kTAM)* augments the aTAM with a stochastic model of self-assembly dynamics [6], yielding a more realistic model of molecular self-assembly. The dynamics of kTAM allow tiles to attach to each other in violation of the attachment rules of aTAM. This can result in assembly errors, *i.e.*, departures

---

* This work was supported in part by a research grant from the Natural Sciences and Engineering Research Council (NSERC) of Canada.

R. Deaton and A. Suyama (Eds.): DNA 15, LNCS 5877, pp. 116–125, 2009.

from ideal growth of the assembly. One means of controlling such assembly errors is through *proofreading tile sets*, constructions of which have been suggested by Winfree and Bekbolatov [10], Chen and Goel [1], Reif *et al.* [4], and Soloveichik and Winfree [6]. Each of these constructions has its strengths and weaknesses, as discussed in [6].

In this paper, we consider a completely different type of error, one arising not due to imperfections in the process of assembly, but rather due to imperfections in the process of tile set fabrication. We assume a scenario in which, at the time of fabricating tiles out of DNA molecules, some small proportion of the tiles is imperfectly created. This could happen, for example, due to the denaturing of molecules caused by chemical or environmental effects, or simply due to human error in the fabrication process. We use the term *malformed tiles* to denote these imperfectly created tiles. We consider a model of malformed tile creation in which some of the tile edges receive labels different from the designed (correct) labels. We then ask the following question: how do tile sets containing a small proportion of malformed tiles deviate from their designed assembly behaviour under kTAM? To the best of our knowledge, the behaviour of tile assemblies in the presence of malformed tiles has not been previously studied in the literature.

We present extensive simulation results comparing error rates in assemblies with and without malformed tiles. In our simulations, carried out within Winfree's tile assembly simulator, xgrow [11], we considered a range of different tile sets, including several different constructions of proofreading tile sets. As expected, assembly-error rates in tile sets with malformed tiles exceeded those in tile sets without such tiles. Also predictably, the extent of resilience to malformed tile errors varies from one tile set to another. Most notably, the snaked proofreading tile set of Chen and Goel [1] performed very poorly in simulations. In the presence of malformed tiles, this tile set showed a tendency to fatally stall, failing to even form moderately sized tile aggregates[1], While looking for an explanation for this phenomenon, we discovered two new snaked proofreading tile set constructions that resist the effect of malformed tiles quite well. These new constructions constitute a significant contribution of this paper.

The rest of the paper is organized as follows. In Section 2, we provide brief descriptions of the tile assembly models aTAM and kTAM, and of the various proofreading tile set constructions we consider in this paper. Section 3 describes our model for malformed tiles, Section 4 contains our simulation results, and Section 5 discusses malformed snaked proofreading tile sets. We make some concluding remarks in Section 6.

2   Background

We provide here a brief description of Winfree's tile assembly models sufficient for our purposes; for more details, the reader may refer to [1],[4],[5], or [9].

2.1   The Tile Assembly Models aTAM and kTAM

The basic building block of aTAM is a square *tile*, which has four edges, and which belongs to one of finitely many *tile types*. The collection of tile types allowed in the model is called the *tile set*. Associated with each tile is a *tile label* taken from some

---

[1] In this paper, we use the terms "aggregate" and "aggregation" to mean an assembly of tiles.

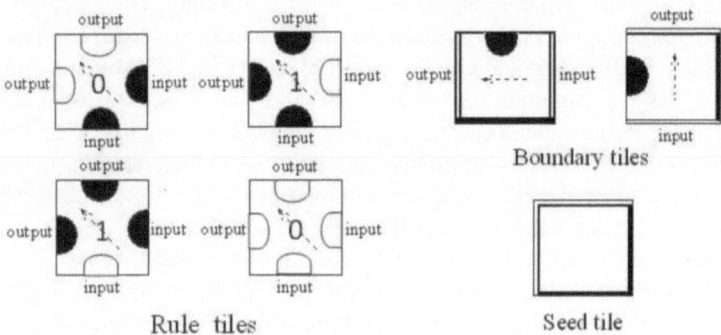

Fig. 1. The basic Sierpinski tile set. Different edge labels are represented by different depictions of edges: lines with a white or black semicircle, double lines and thick solid lines. An edge with a black or white semicircle has glue strength 1, a double line has glue strength 2, while a thick solid line has glue strength 0. Tile labels (0/1) have been placed at the centre of each tile.

finite alphabet. The edges of tiles are given *edge labels*, and furthermore, each edge label has a non-negative integer called the *glue strength* associated with it. The tile label, edge labels, and glue strengths for a tile are completely determined by the tile type to which it belongs. Two edges from distinct tiles are called *matching edges* when they both have the same edge label. Matching edges can attach to each other to form *bonds*, the strength of the bond being equal to the glue strength of either edge in the matching pair forming the bond. Bonds of strength 0, 1, and at least 2, are called *null*, *weak* and *strong* bonds, respectively. Assembly occurs by the iterative addition of tiles to an existing aggregate, initialized by a special *seed tile*. In aTAM, a tile can be legally added to a position at the border of an existing aggregate if the total strength of the bonds formed by the matching edges in that addition is greater than or equal to a system parameter called *temperature*, which is always a positive integer. The arrangement of tile labels obtained from an assembly of tiles is called a *pattern*.

Figure 1 shows an example of the *basic Sierpinski tile set*. This tile set consists of four kinds of *rule tiles*, two kinds of *boundary tiles*, and one seed tile. If the system temperature is set to 2, which means that each addition of a new tile to an existing aggregate requires at least two weak bonds or one strong bond, then upon initialization by the seed tile, the Sierpinski tile set forms the Sierpinski *pattern* (see *e.g.*, [4, Figure 1] or [6, Figure 1]). This assembly consists of a "reverse-L" shaped frame consisting of the seed tile at the corner and boundary tiles along the arms of the reverse L, to which the rule tiles attach forming the internal part of the assembly. The assembly grows in a bottom-right to top-left fashion, determining an orientation on the tiles in the tile set. The orientation of a tile is from its *input sides* to its *output sides*, as depicted by the dashed arrows in Figure 1. A boundary tile has only one input side and one output side. The basic Binary Counter tile set [9] is another simple tile set which has the same number of rule and boundary tiles as the Sierpinski tile set.

The kTAM is an augmentation of aTAM that includes a *forward rate* for tiles to associate to the growing assembly, and a *reverse rate* for tiles to detach themselves from the assembly. The forward rate of attachment of a tile depends only on the overall concentration of tiles in the ambient solution, and is given by $k_f = ke^{-G_{mc}}$, where G_{mc} is the negative logarithm of tile concentration, and k is a constant that sets the time scale. The rate at which a tile dissociates from the assembly depends on how strongly the tile is attached to the assembly. The dissociation rate is given by $k_{r,b} = ke^{-bG_{se}}$, where k is the same constant as above, b is the total strength of the bonds between the tile and the assembly, and G_{se} denotes the free energy of breaking a single bond. The ratio $\frac{G_{mc}}{G_{se}}$ plays a role in kTAM similar to that of the system temperature in aTAM. As a result, an attachment that is legal in aTAM is also favoured to happen in kTAM.

However, it can happen within kTAM (but not within aTAM) that a tile attaches to the perimeter of the aggregate with insufficient bond strength, but before it falls off, another tile attaches next to it as a legal addition, resulting in a stable assembly overall. The initial insufficient attachment may be due to mismatched edges (*growth error*), or may be due to the attachment of an isolated tile at a facet of the aggregate (*facet error*). Such errors remain "frozen" within the assembly, and may cause further errors, resulting in a final assembly that differs from the designed (correct) final assembly.

Recall that the arrangement of tile labels in an aggregate forms a pattern; a pattern corresponding to an error-free aggregate is a *correct pattern*. The arrangement of tile labels in an aggregate that contains errors may still form the correct (designed) pattern, but it is more likely that such an aggregate results in a *wrong pattern*, *i.e.*, an arrangment of tile labels that is different from the designed pattern.

2.2 Proofreading Tile Sets

Proofreading tile sets are a means of controlling errors in the tile assembly process. Loosely speaking, the idea behind such schemes is to start with a basic tile set (such as the Sierpinski tile set in Figure 1), and modify it by introducing redundant information into the edge labels of tiles, which can then act as an error-control mechanism. The first example of such a tile set construction was due to Winfree and Bekbolatov [10], who transformed each tile of the basic tile set into a larger $k \times k$ block of proofreading tiles. However, the overall error rate for this scheme did not scale well with k, because such proofreading tile sets, while being able to control growth errors, were largely ineffective against facet errors. In [1], Chen and Goel introduced the $k \times k$ *snaked proofreading tile set* that modified the Winfree-Bekbolatov construction by changing the glue strengths on some of the edges of the constituent tiles within each $k \times k$ block. By appropriately introducing strength-0 and strength-2 edges into the constituent tiles, they were able to control the manner in which the constituent tiles assembled to form a $k \times k$ block. This allowed them to successfully deal with facet errors (as well as growth errors), and they were able to show that error rates in their assemblies decreased exponentially with k.

Both the constructions mentioned above reduce error rates at the cost of scaling up the size of the final assembly by a factor of k^2. Reif *et al.* [4] showed that certain basic tile sets (such as the Sierpinski and Binary Counter tile sets) could be transformed into *compact* proofreading tile sets that lowered error rates without scaling up the size of the final assembly. They gave two specific constructions — a "2-way overlay" construction and a "3-way overlay" construction — in which each edge label in a proofreading tile encoded two or three bits of partially redundant information. In [6, Appendix B],

Soloveichik and Winfree give a (different) construction, which they attribute to Paul Rothemund and Matthew Cook, of a general "k-way overlay" compact proofreading tile set. It should again be stressed that, while the number of tiles in any of these compact proofreading tile sets is larger than that in the original Sierpinski or Binary Counter tile set, the size of the final assembly required to produce an $N \times N$ Sierpinski or Binary Counter pattern is exactly the same as that for the original tile set.

3   Malformed Tile Sets

As stated in the introduction, the object of this paper is to consider a problem different from the one tackled by proofreading tile set constructions. The scenario that motivates us is one in which errors affect the process of tile set fabrication, resulting in abnormalities or imperfections in some small proportion of the tiles. Thus, in our set-up, malformed tiles are simply deformations of normal tiles. In fact, since the only sort of tile abnormalities that influence the assembly process are imperfections in tile edges, we will assume that abnormal tiles are created by modifying the edges of normal tiles.

So, suppose that we are given a *normal tile set* $\mathfrak{T}$ consisting of tile types that we consider to be *normal tile types*. A *malformed tile type* is obtained by modifying some (or all) of the edge labels of some normal tile type. The edges in the malformed tile type whose labels differ from the normal edge labels will be called *malformed edges*. A *malformed tile set* derived from $\mathfrak{T}$ is a collection of tile types consisting of the (normal) tile types in $\mathfrak{T}$ along with some malformed tile types derived from the tile types in $\mathfrak{T}$.

We now describe certain simplifying assumptions we make in our model for malformed tile sets. Our first assumption is based on an understanding that edge imperfections do not transform one normal edge label into another normal edge label.

Assumption 1. *Malformed edges always receive labels that are distinct from all the edge labels of normal tile types.*

We do not specify at this point whether or not distinct malformed edges can receive the same edge label. There are two extreme cases that could be considered — one in which no pair of distinct malformed edges receives the same edge label, and another in which all malformed edges receive identical edge labels. We only focus on the former case in this paper; the latter case receives some attention in [3].

The next assumption is consistent with the fact that malformed tile types are obtained from normal tile types only through edge modifications.

Assumption 2. *A malformed tile type has the same tile label as the normal tile type from which it is formed.*

Note that this assumption allows for the possibility that a tile assembly based on a malformed tile set may contain mismatched edges and still produce the same (correct) pattern as an assembly obtained from the normal tile set.

Assumption 3. *A malformed tile type contains exactly one malformed edge.*

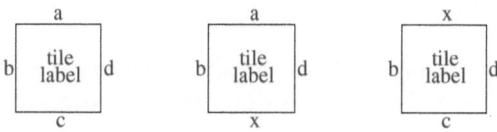

Fig. 2. The tile on the left is a normal rule tile, the tile in the middle is an IM tile type, and the tile on the right is an OM tile type

The rationale behind Assumption 3 is that, among malformed tiles, those with more than one malformed edge have a negligible impact on the assembly process, as they are unlikely to form stable attachments with an existing aggregate.

The definitions and assumptions we have made up to this point remain valid for an arbitrary normal tile set. However, to make our subsequent discussions easier, we will from now on restrict ourselves only to the normal tile sets covered by our simulations, namely, the basic Sierpinski tile set (Figure 1), the basic Binary Counter tile set, and various proofreading tile sets derived from these basic tile sets. There are three kinds of tile types in these tile sets — a unique seed tile, boundary tiles, and rule tiles.

We will assume that only rule tiles can be modified to form malformed tiles. This is because malformed edges on the seed tile or on boundary tiles either do not influence the tile assembly process at all or cause the process to come to a standstill. To illustrate this more clearly, we take the example of the Sierpinski tile set in Figure 1. Note that imperfections in the strength-0 edges of the seed or boundary tiles would have no influence on the tile assembly process. On the other hand, a malformed strength-2 edge in a seed tile or boundary tile would make the formation of a stable assembly highly unlikely. Boundary tiles in this tile set also have strength-1 edges, imperfections in which would have an influence on the assembly process. But the effect of such a malformed edge can be equivalently modelled by modifying the matching strength-1 edge in the rule tile that must attach to it. We are thus justified in making the following assumption.

Assumption 4. *Malformed tiles are obtained by modifying rule tiles only.*

Rule tiles have two input edges (bottom and right) and two output edges (top and left). To differentiate between the effects of malformed input edges and malformed output edges, we categorize malformed tile sets into three classes:

(1) *Input-side malformed (IM) tile sets*, in which the malformed tile types consist of all tile types obtainable by modifying one input edge of a normal rule tile.
(2) *Output-side malformed (OM) tile sets*, in which the malformed tile types consist of all tile types obtainable by modifying one output edge of a normal rule tile.
(3) *Input/output-side malformed (IOM) tile sets*, in which the malformed tile types consist of all possible IM tile types and OM tile types.

Figure 2 contains a depiction of these malformed tile types. In our simulations, we deal separately with each of the above classes of malformed tile sets.

4 Simulation Results

The simulations we performed provide a broad comparison of the performance, in the presence of malformed tiles, of various kinds of tile sets producing either the Sierpinski pattern or the Binary Counter pattern. Due to space constraints, we only report in detail here the simulation results for Sierpinski tile sets; the results for Binary Counter tile sets are similar, and can be found in [3].

We report here simulation results for the basic Sierpinski tile set (Figure 1), and the following proofreading tile sets derived from it: the Winfree-Bekbolatov 2×2 proofreading tile set [10], the 2- and 3-way overlay tile sets of Reif *et al.* [4], and the Rothemund-Cook k-way overlay tile sets [6, Appendix B]. We also ran simulations on the Chen-Goel snaked proofreading tile set [1], and found that this tile set performs very poorly in the presence of malformed tiles, as we explain in more detail in Section 5.

All simulations were performed within the xgrow simulator [11]. The values of G_{mc} and G_{se} were set to 17 and 8.6, respectively. For each type of rule tile, the tile concentration value was set to [0.9]; for each malformed tile type, the tile concentration value was set to [0.05]; the seed tile concentration was [0.1], and the concentration of each boundary tile type was [0.3]. We ran 50 simulations for each tile set. All our malformed tile sets have the property that malformed edges in distinct malformed tile types receive distinct edge labels (see the remarks following Assumption 1).

As one would expect, malformed tile sets perform worse in simulations than normal tile sets. Furthermore, tile sets containing OM tile types perform worse than IM tile sets, in the sense that aggregations formed by the former contain more malformed tiles, and hence more mismatched-edge errors (as reported by xgrow), than aggregations formed by the latter. This phenomenon has a straightforward explanation within the kTAM model. Consider any open position in the growing asembly where a normal tile can stably attach itself (by forming two weak bonds). An OM tile derived from the same normal tile has an equal chance of attaching itself to that position, as the input edges of the malformed tile are the same as those of the normal tile. On the other hand, an IM tile has one malformed input edge, so its attachment to the open position must happen with insufficient bond strength, which makes the attachment less likely to occur. This explains why the proportion of malformed tiles attaching to an aggregation in an OM or IOM tile set is higher than that in an IM tile set.

The influence of malformed tiles on various proofreading tile set constructions for the Sierpinski pattern is tabulated in Table 1. The entries in the table attempt to measure the ability of several different malformed tile sets to form a 128×128 block of the Sierpinski pattern. In the case of malformed 2×2 proofreading tile sets, this requires the formation of a 256×256 aggregate of tiles; but in all other cases, only a 128×128 tile aggregate is required. The first number in each table entry records the proportion, out of a total of 50 aggregations, of aggregations that contain mismatched-edge errors. The second number (the one in parentheses) represents the proportion, out of 50, of aggregations that result in a wrong 128×128 pattern. Recall that, by virtue of Assumption 2, it is possible for an assembly formed from malformed tile sets to produce the correct pattern even if the assembly contains mismatched edges. Thus, the second number in any of the entries in the table can never exceed the first number in that entry; in fact, the second number is often significantly smaller.

Table 1. A comparison of various malformed proofreading tile sets for the Sierpinski pattern. The first number in each table entry is the proportion, out of 50 tile aggregations, of aggregations that contain mismatched-edge errors. The second number, in parentheses, is the proportion of aggregations that fail to produce a correct 128×128 block of the Sierpinski pattern.

	Normal tile set	IM tile set	OM tile set	IOM tile set
Basic Sierpinski tile set	1 (1)	1 (1)	1 (1)	1 (1)
2×2 proofreading tile set	0 (0)	0.56 (0.02)	0.76 (0.44)	0.94 (0.48)
Reif *et al.* 2-way overlay	0 (0)	0.3 (0.06)	0.42 (0.14)	0.58 (0.2)
Reif *et al.* 3-way overlay	0 (0)	0.18 (0)	0.24 (0.04)	0.30 (0.04)
Rothemund-Cook 1-way overlay	0 (0)	0.26 (0.06)	0.34 (0.1)	0.58 (0.16)
Rothemund-Cook 2-way overlay	0 (0)	0.36 (0)	0.38 (0.02)	0.52 (0)
Rothemund-Cook 3-way overlay	0 (0)	0.26 (0)	0.44 (0.02)	0.56 (0.02)

Table 2. Performance of IOM tile sets derived from the Rothemund-Cook k-way overlay tile sets

	$k = 1$	$k = 2$	$k = 3$	$k = 4$
Proportion of 256×256 aggregates with errors	0.98	0.96	0.92	0.94
Proportion of 256×256 aggregates forming wrong patterns	0.84	0.4	0.22	0.32

Among the malformed tile sets considered in Table 1, those derived from the 3-way overlay tile set of Reif *et al.* produce the most number of error-free aggregations. Interestingly, the malformed Rothemund-Cook 2-way and 3-way overlay tile sets are remarkably consistent in producing correct 128×128 blocks of the Sierpinksi pattern almost all the time. This observation induced us to study the performance of these tile sets over a larger assembly size of 256×256 tiles. We restricted our attention to the worst-case IOM tile sets only. Fifty (50) simulation runs were performed on each of the IOM tile sets derived from the Rothemund-Cook k-way overlay construction, for $k = 1, 2, 3, 4$. The simulation results, presented in Table 2, show that despite the fact that almost all the 256×256 tile aggregates produced by these IOM tile sets have mismatched-edge errors in them, the proportion of aggregates that fail to produce the correct 256×256 Sierpinski pattern block is (at least for $k = 2, 3, 4$) much smaller.

Note from Table 2 that the malformed k-way overlay tile sets improve in performance as k increases from 1 to 3, but the improvement stops there. Two things happen in these tile sets as k increases — the amount of redundant information encoded in each normal edge label increases, and also, the total number of normal (and hence, malformed) tile types increases. These two factors seem to affect performance in opposite ways. Increased redundancy in the edge labels appears to mitigate, to some extent, the effect of malformed tiles, while an increase in the number of tile types seems to adversely affect performance. Indeed, evidence can be found even in Table 1 for the fact that the number of normal tile types plays a role in determining the performance of derived malformed tile sets. The 2-way overlay tile set of Reif *et al.*, and the Rothemund-Cook 1-way overlay tile set have eight (normal) tiles each, and the malformed tile sets derived from them display remarkably similar performance.

5   Malformed Snaked Proofreading Tile Sets

We observed in the previous section that malformed tile sets derived from the snaked proofreading tile sets of Chen and Goel [1] performed very poorly in our simulations. The simulations we undertook focused on malformed tile sets obtained from the 4×4 snaked proofreading tile set generating the Sierpinski pattern. On the one hand, IM tile sets were able to form 256×256 aggregations, but these aggregations were riddled with mismatched-edge errors. On the other hand, OM and IOM tile sets failed to even produce aggregations of this size. In our simulations, OM tiles caused the tile assembly process to stall repeatedly and frequently. The slowdown caused by stalling was so extreme that even when our simulations were run without interruption for several hours, we could not get an OM tile set to produce a complete 256×256 assembly, By way of comparison, OM tile sets derived from the Winfree-Bekbolatov 4×4 proofreading construction were able to form 256×256 aggregations in a matter of minutes.

In trying to understand the reason for the stalling phenomenon described above, we noticed that stalling almost always occurred in situations when a complete 4×4 block was formed, but there were OM tiles frozen within the output-side (top and left) edges of the complete block. The frozen OM tiles hindered the formation of the next 4×4 block, and it took an exceedingly long time for them to fall off according to the dynamics of kTAM. The reason for this is that, as noted in [1], all the tiles on the output sides of the 4×4 block in Figure 3(a) are held by bonds of total strength at least 3. So whenever all the tiles on a block are attached, it is unlikely for them to fall off, leaving any OM tiles (with malformed edges facing outside) frozen within the assembled block.

To get around this problem, we propose two new snaked proofreading constructions, referred to as Construction B and Construction C, which are depicted in Figures 3(b) and 3(c), respectively. Note that, in each of these constructions, the last tile to attach within a block (identifiable by the circle in the corresponding figure) is held to the assembly by bonds of total strength 2. Thus, if the assembly were to stall after a complete 4×4 block is formed, the chances of the last tile falling off are higher in these constructions than in the Chen-Goel construction. Furthermore, the new constructions (essentially) preserve the "snaked" growth order of the Chen-Goel construction, which was shown in [1] to be effective in combating the influence of growth and facet errors.

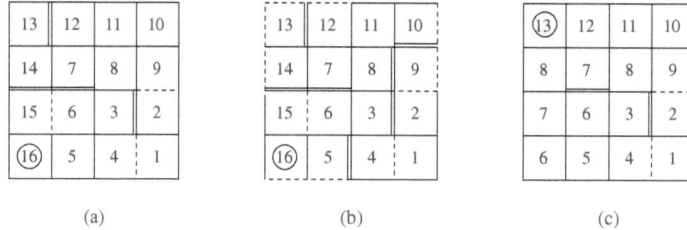

Fig. 3. Snaked proofreading constructions: (a) the Chen-Goel construction, (b) Construction B, (c) Construction C. Dashed edges, single edges, and double edges depict strength-0, strength-1, and strength-2 bonds, respectively. The numbering of the tiles shows the order in which tiles attach to form the 4×4 block. The circles identify the last tiles to attach in a block.

The table below shows simulation results for OM tile sets derived from each of the constructions in Figure 3. Fifty simulation runs were performed on each OM tile set, with a target assembly size of 256×256 tiles. Simulations were run in batches of 25 parallel runs on a standard desktop computer, with each batch allowed to run uninterrupted for 30 hours. The first number in each table entry is the number of runs, out of 50, that resulted in a complete 256×256 tile assembly, and the second number (within parentheses) is the number of runs that resulted in a complete and correct 64×64 block of the Sierpinski pattern.

Chen-Goel	Construction B	Construction C
0 (0)	30 (29)	36 (33)

It is clear that the two new snaked proofreading constructions are reasonably resilient to malformed tile errors, and vastly outperform the Chen-Goel construction.

6 Conclusion

In this paper, we studied, through xgrow simulations, the effect of malformed (*i.e.*, imperfectly-fabricated) tiles on the performance of various Sierpinski tile sets. A significant contribution of this paper is the development of two new snaked proofreading tile set constructions that provide examples of tile sets that are robust with respect to errors intrinsic to the assembly process, and also with respect to malformed tiles.

It should be pointed out that in this paper, we restricted our attention to proofreading tile sets in which the error-control mechanism consists of the introduction of redundancy into the edge labels of tiles. Alternative error suppression mechanisms have indeed been proposed in the literature (for example, [2]), and it would be of interest to study how malformed tiles affect these kinds of proofreading mechanisms.

References

1. Chen, H., Goel, A.: Error Free Self-assembly Using Error Prone Tiles. In: Ferretti, C., Mauri, G., Zandron, C. (eds.) DNA 2004. LNCS, vol. 3384, pp. 62–75. Springer, Heidelberg (2005)
2. Fujibayashi, K., Zhang, D.Y., Winfree, E., Murata, S.: Error Suppression Mechanisms for DNA Tile Self-Assembly and their Simulation. Natural Computing (July 2008)
3. Meng, Y.: MSc thesis (in preparation)
4. Reif, J.H., Sahu, S., Yin, P.: Compact Error-Resilient Computational DNA Tiling Assemblies. In: Ferretti, C., Mauri, G., Zandron, C. (eds.) DNA 2004. LNCS, vol. 3384, pp. 293–307. Springer, Heidelberg (2005)
5. Rothemund, P.W.K., Papadakis, N., Winfree, E.: Algorithmic Self-Assembly of DNA Sierpinski Triangles. PLoS Biology 2(12), 2:e424 (2004)
6. Soloveichik, D., Winfree, E.: Complexity of Compact Proofreading for Self-Assembled Patterns. In: Carbone, A., Pierce, N.A. (eds.) DNA 2005. LNCS, vol. 3892, pp. 305–324. Springer, Heidelberg (2006)
7. Wang, H.: Proving Theorems by Pattern Recognition II. Bell Syst. Tech. J. 40, 1–42 (1961)
8. Winfree, E.: Algorithmic Self-Assembly of DNA. PhD thesis, Caltech (1998)
9. Winfree, E.: Simulation of Computing by Self-Assembly. Caltech Technical CS Report (1998)
10. Winfree, E., Bekbolatov, R.: Proofreading Tile Sets: Error Correction for Algorithmic Self-Assembly. In: Chen, J., Reif, J.H. (eds.) DNA 2003. LNCS, vol. 2943, pp. 126–144. Springer, Heidelberg (2004)
11. The 'xgrow' simulator, http://www.dna.caltech.edu/Xgrow

Positional State Representation and Its Transition Control for Photonic DNA Automaton

Hiroto Sakai, Yusuke Ogura, and Jun Tanida

Department of Information and Physical Sciences, Graduate School of Information
Science and Technology, Osaka University
1-5 Yamadaoka, Suita, Osaka 565-0871, Japan
{h-sakai,ogura,tanida}@ist.osaka-u.ac.jp

Abstract. DNA automaton is utilized as a processing core of a nano-scale processor that can deal with molecular information directly. We are studying on photonic DNA automaton in which light and DNA are used as information carriers. In the automaton, the internal states are represented as the conformation of DNA, or the spatial position of the specific bases in the DNA strand, and the state-transition is induced by an external optical signal. This paper focuses on a method for state representation and state transition control based on the conformation of DNA. An implementation using hairpin-DNA is experimentally demonstrated to verify the proposed scheme.

1 Introduction

DNA automaton is an example of implementation of computation using DNA. Automaton is a fundamental model of computation, and it is useful in various systems for information processing. A complicated processing is possible as a sequence of transitions of the internal state of the automaton, which depend on input and a current internal state. E. Shapiro *et al.* demonstrated programed control of release of a single-stranded DNA that works as a drug using automaton[1].

Combining another carrier of information with DNA is an effective strategy to utilize the capability of DNA computing. Light is a powerful information carrier because information that is encoded into optical signals is carried in parallel. Flexible manipulation of the light is achievable using optical devices such as lenses, optical fibers, spatial light modulators, and a variety of optical field distributions can be generated easily. Optical techniques offer local, remote, contact-less accessibility and controllability on various objects including molecules, and make it possible to broadcast the information to the objects. We are studying about photonic DNA automaton, which utilizes light and DNA as information carriers[2,3]. In a previous paper, we proposed a method for realizing photonic DNA automaton using change of the conformation of DNA depending on photonic signals. We designed four reaction schemes, which produce

R. Deaton and A. Suyama (Eds.): DNA 15, LNCS 5877, pp. 126–136, 2009.

different conformations of the DNA depending on a binary photonic signal, and demonstrated the schemes experimentally[3]. To implement our photonic DNA automaton, the conformation of the DNA has to be converted to the position of a DNA strand which induces the following reaction at a local position. The position of the DNA is utilized to represent the internal state of the automaton. The function of the automaton is regulated by repetition of a conformation-change and the following position-change of DNA.

In this paper, we report on a method for controlling the position of a DNA strand in a DNA nano-structure using conformation-change of the DNA to control a following reaction at a local position. The method enables us to use positional information of the DNA, which is converted from information of DNA-conformation. Various structures of DNA have been designed and demonstrated using self-assembly[4]. We use 4-arm DNA junctions and 3-arm junctions to fabricate a rectangular DNA structure as a basic working place where the behavior of the photonic DNA automaton is controlled. Hairpin-DNAs are incorporated on edges of the rectangular structure, and the conformation of the hairpin-DNAs on a single edge is converted to the positional information. We experimentally confirmed the position-change of DNA using conformation-change and the following reaction of the hairpin-DNAs dependent on the position of the DNA using two rectangle structures, which shared an edge. Experimental results demonstrate a few behaviors necessary for state-transition, which is realized by change of the position of the DNA and the following conformation-change of the hairpin-DNAs in the rectangle structures. Combining reaction schemes on these partial behaviors with that depending on photonic signals, we can implement a photonic DNA automaton with two symbols and two states. Section 2 explains the concept of photonic DNA automaton. In Section 3, we describe an implementation method of the automaton and a method for controlling the position of DNA using a pair of hairpin-DNAs. In Section 4, we show experimental results about a part of the function of the automaton. In Section 5, we give conclusions.

2   Photonic DNA Automaton

Figure 1 shows the concept of photonic DNA automaton, which is an implementation of automaton using photonics and DNA. The states are represented by DNA, and the rules of state-transition are determined by DNA reaction systems assigned to the individual photonic DNA automata. Various light field distributions can be generated, and information is transmitted to automata by signal broadcasting through light. The photonic DNA automata work in parallel by following the optical signals. Although the resolution of the optical signals is a micrometer scale, the effective resolution of information dealt with light achieves a nanometer scale because photonic DNA automata perform computations in a nanometer scale. When controlling DNA reactions by photonics, we have some difficulties to suppress unexpected progress of reactions using only light-irradiation. We study a method for controlling reactions of DNA in stepwise using light.

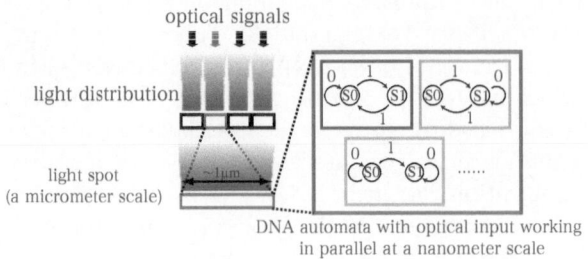

Fig. 1. The concept of photonic DNA automaton

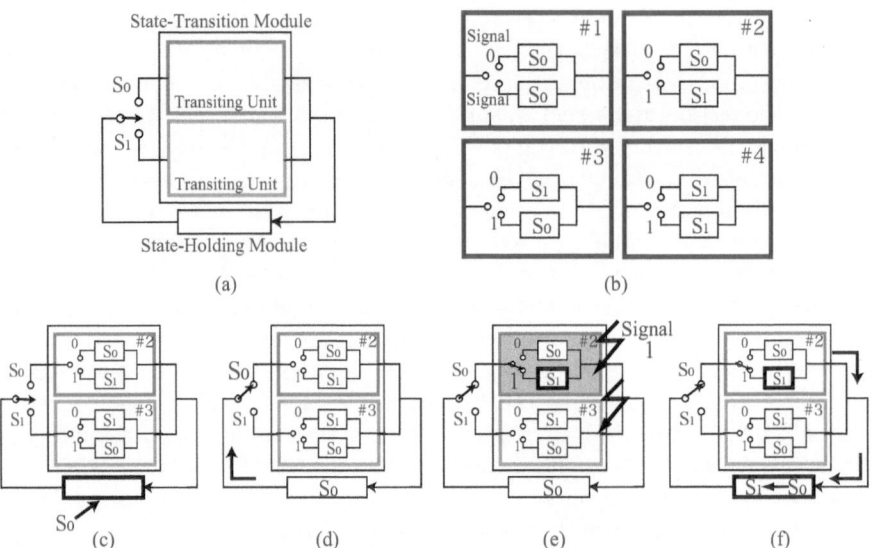

Fig. 2. The implementing method for photonic DNA automaton. (a) The overview of the system, (b) The four schemes for representing transiting rules of an automaton, (c) Setting of the initial state to the SHM, (d) Selecting of transiting unit depending on the initial state, (e) Transiting of the current state to next state in response to a signal, (f) Renewing of the current state to next state.

Figure 2 (a) shows a system of photonic DNA automaton. The system consists of two modules: the state-transition module (STM) and the state-holding module (SHM). The STM receives optical signals and executes state-transitions in response to the signals. The SHM temporarily stores a current state, which is renewed after receiving information on the next state from the STM. These modules are actuated alternately. Let us consider an automaton with two internal-states, S_0 and S_1, and two input symbols, 0 and 1. The STM is constructed with two transiting units that correspond to each of the internal states, S_0 and S_1. The transiting units are used to determine the next state by receiving a photonic signal. One of the four schemes, which are used for four possible state-transiting patterns, is assigned to each of the transiting units. The four transition rules considered to implement the automaton are shown in Fig. 2 (b). These schemes implement the transiting rules depending on only an input symbol.

Figures 2 (c) through (f) show a procedure of operating an automaton. At first, an initial-state is loaded to the SHM (Fig. 2 (c)). The information on the current state is transferred from the SHM to the STM by selecting one of the transiting units depending on the state (Fig. 2 (d)). Only the selected transiting unit works in the following process. The state is transited to the next one by following the input signal (Fig. 2 (e)). The information on the next state is transferred from the STM to the SHM. In the SHM, the state is renewed to the next state (Fig. 2 (f)). The behavior of the photonic DNA automaton is regulated by repeating these operations. The system is possible to implement arbitrary state-transition diagrams with two states and two kind of input symbols by selecting the scheme used for each of the transiting units among four schemes.

3   The Implementation Method of an Automaton

In the implementation method, the state of photonic DNA automaton is represented as the positional information of a strand in a DNA nano-structure. To deal with the positional information, a pair of hairpin-DNAs are used. In this paper, we describe on conversion of conformation of the hairpin DNAs to positional information for state-transitions.

Hairpin-DNAs can form two different stable structures: an open-state and a closed-state. Figure 3 shows switching between two states by adding two kinds of DNA strands. Let assume that the closed-state hairpin-DNA consists of a loop, a stem, and a sticky-end. When the sequence of the sticky-end and the neighboring stem is complement to that of a liner strand L_1, these strands anneal together, and the hairpin-DNA becomes the open-state as the result of branch-migration. If strand L_1 includes the sequence which is complement to another strand, L_2, two strands can anneal each other. L_1 is removed from the open-state hairpin-DNA, and it returns to the closed-state. In this paper, strands L_1 and L_2 are called an opener DNA and a closer DNA, respectively.

The scheme of converting conformation of a DNA nano-structure to the position of a specific DNA strand in the DNA structure is shown in Fig. 4. A solid

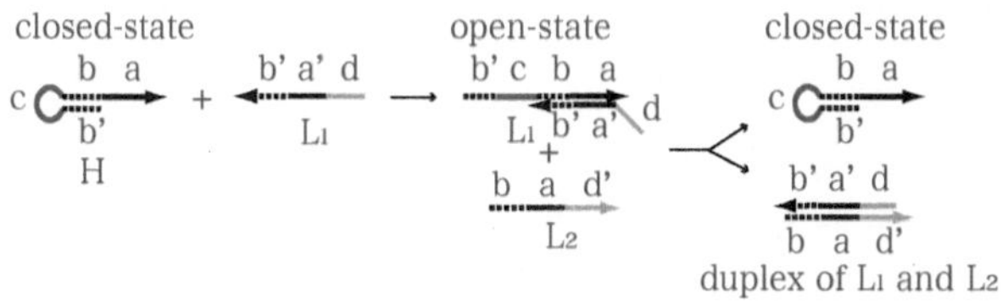

Fig. 3. Switching of conformations of a hairpin-DNA using two kind of DNA strands

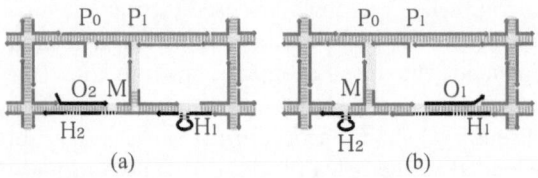

Fig. 4. The scheme of converting conformation of DNA to position of specific bases of the DNA. (a) H_1 is closed; H_2 is open; and M is located at P_1, (b) H_1 is open; H_2 is closed; and M is located at P_0.

line indicates a single-stranded DNA. A rectangular structure made of a number of single-stranded DNAs is used as a working-basis. A pair of hairpin-DNAs are incorporated into the working-basis. When a hairpin-DNA, H_1, is closed and the other hairpin-DNA, H_2, is open, a liner strand, M, arranged between H_1 and H_2 is located at position P_1 (Fig. 4 (a)). By adding an opener DNA, O_1, which opens H_1 and a closer DNA, C_2, which closes H_2 to the solution of the working-basis, strand M moves to position P_0 (Fig. 4 (b)). The position of M can be returned back to the initial position, P_1, by adding other strands that close H_1 and open H_2 (Fig. 4 (a)). By repeating these operations, the position is changed reversibly. This operation is used for transforming the conformation of the pair of the hairpin-DNAs into positional information of strand M. When the upper edge of the working-basis contains two sticky-ends, which are the complement sequence to the sticky-end of M, M can react with only one of the sticky-ends in the upper edge depending on its position.

Figure 5 shows the operating procedure of an automaton. The working-basis consists of two rectangular-shape structures which share an edge as shown in Fig. 5(a). The top, middle, and bottom tiers in the structure work as the SHM for setting up of the initial states (Fig. 2 (c)), the STM (Fig. 2 (e)), and the SHM for renewing to the next state (Fig. 2 (f)), respectively. The conformations of hairpin-DNAs, B_1 and B_2, on the side edges of the working-basis are controlled using a specific opener and a closer DNA. Transiting units (TUs) for the states S_0 and S_1 on the left of M' in the middle tier consist of a hairpin-DNA which reacts with M. Each TU corresponds to one of four reaction schemes, in which conformation of the hairpin-DNAs is changed in response to the optical signal, 0 or 1 (Fig. 5 (b)).

We demonstrated photonic control of conformational change of the hairpin-DNAs which were tethered with azobenzene[3]. The binding force between a DNA strand with trans-form azobenzene and its complementary DNA strand is stronger than that with cis-form azobenzene. Because azobenzene is switched to trans- or cis-form using visible and ultraviolet light, the binding of the DNA strands can be controlled in response to light at an adequate range of temperature. The more azobenzenes are attached into each DNA, the more efficient control of a duplex is possible although the transformation efficiency of individual azobenzenes is low[5]. Strand M corresponds to M_0 in TUs, and can react

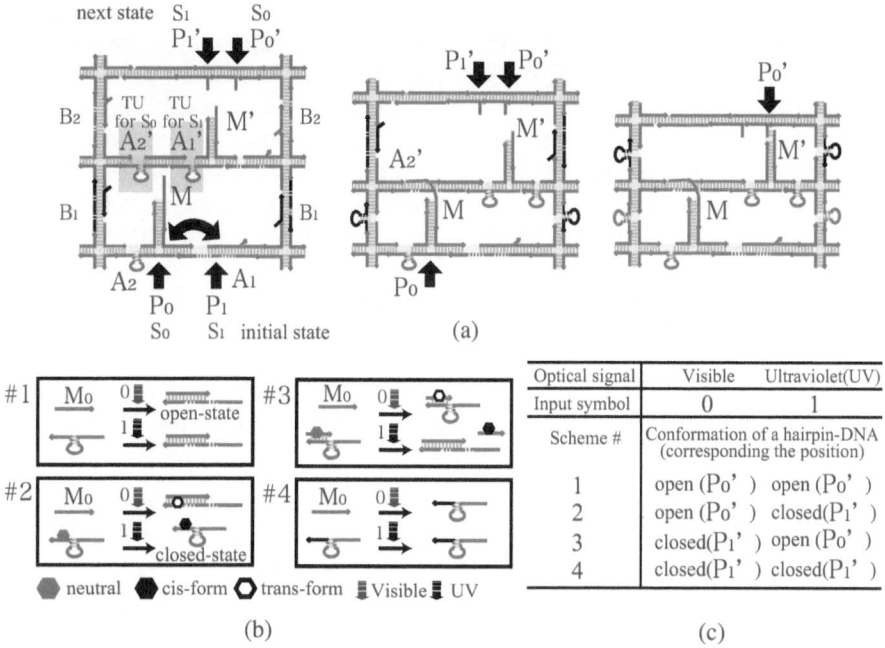

Fig. 5. The operating procedure of an automaton. (a)Left; The composition of an automaton using working-bases (setting an initial state on the bottom tier (SHM)), Middle; transiting state on the middle tier (STM), Right; holding the next state (SHM). (b) Optical modules for representing transiting rules of an automaton. (c) The relationship between input symbols encoded into light and the corresponding position that changes according to the conformation of a hairpin-DNA.

with one of two hairpin-DNAs, A_1' and A_2', which work as the transiting units prepared for the individual internal states. Note that when hairpin-DNAs, B_1, are the open-state as shown in the left figure of Fig. 5 (a), M can react with neither A_1' nor A_2' because M on the bottom tier doesn't reach to the middle tier. Using strands B_1 and B_2, the transferring of the positional information from the bottom/middle tier to the middle/top tier is regulated in stepwise.

Let us consider a cycle of state-transition of an automaton with two kinds of input symbols and two internal-states. Figure 5 (a) is illustrated with assuming that both of the TUs correspond to reaction scheme $\sharp$ 1 as an example. First, an initial state, S_0 or S_1, is set on the bottom tier (Fig. 5 (a)Left). Strand M is horizontally moved to a position, P_0 or P_1, corresponding to the initial state using conformation-change of two hairpin-DNAs A_1 and A_2 in the bottom tier. Second, an optical signal (visible light for signal 0 or UV light for signal 1) is input. This operation determines the position of M' on the middle tier after reaction with strand M in TUs. Third, strand M is vertically moved toward the middle tier by closing hairpin-DNAs B_1 (Fig. 5 (a)Middle). The bottom tier approaches to the middle tier so that single-stranded DNA on each tier can anneal with each other. Only a single TU annealing with strand M starts to work, and the hairpin-DNA in the TU becomes the open- or closed-state by following the optical signal. The position of strand M' changes according to the working TU, which indicates the next state, on the middle tier. If TUs for S_0 and S_1 correspond to ♯ 2 and ♯ 3, conformational change and the following positional change are different in response to the optical signal as shown in Fig. 5 (c).

To hold the information on the next state, the middle tier is moved closer to the top tier by closing hairpin-DNAs B_2 (Fig. 5 (a)Right). The information on the state, which is encoded as the position of M', is transmitted to the top tier by letting M' bind to the sticky-end representing S_0 or S_1 on the top tier. The internal states S_0 and S_1 correspond to positions P_0' and P_1', respectively. Use of the working-basis is effective not only for realizing localized position control but also for achieving faster response based on intra-molecular reactions than inter-molecular reactions. Repeating state-transitions is achievable by increasing rectangular structures in the working-basis.

4   Experiments

In this paper, we focus on state-representation and state-transition of the automaton by controlling the position of a single-stranded DNA using conformation of hairpin-DNAs. In the experiments, we used the reaction of hairpin-DNAs which is independent on light irradiation.

4-arm DNA junctions[6] and 3-arm DNA junctions[7] were used to make a working-basis consisting of a single rectangular structure (Fig. 6(a)) and two rectangular structures with a shared edge (Fig. 6(b)). The working-basis consists of 4-arm junctions, X1 through X4 or X1' through X6', 3-arm junctions Y and Y', and duplexes T and T' with two sticky-ends. In the structures of X, the duplex part of the junction shape has the same sequence shown in references [6] and [7]. The sequences shown in Fig. 6(c) were designed using a software called NUPACK[8].

The first experiment aimed at investigating the behavior of M on the working-basis. Figure 7 shows the conformation controlled using the opener and closer DNAs. To control the conformation of hairpin-DNA H_1 and H_2, two opener DNAs and two closer DNAs were utilized. Strands O_1 and O_2 are the opener DNAs, and C_1 and C_2 are the closer DNAs for H_1 and H_2, respectively. A fluorophore FAM and a quencher BHQ-1 were used as a pair of FRET to detect the position of M. The observed fluorescence intensity depends on the distance between these dyes because excitation energy of FAM is transferred to BHQ-1, and BHQ-1 releases the energy as heat. Strand Y was labeled with BHQ-1 at the 3'-end and a strand T was labeled with FAM at the 5'-end.

At the initial, H_2 is closed and H_1 is open, namely M is at the position P_0 as shown in Fig. 7 (a). When the position of M is P_0, FAM is away from BHQ-1, and high fluorescence intensity is obtained. On the other hand, when M is at P_1, fluorescence from FAM is quenched by BHQ-1 since FAM approaches to BHQ-1. The position (left or right) of M can be determined using a FRET signal.

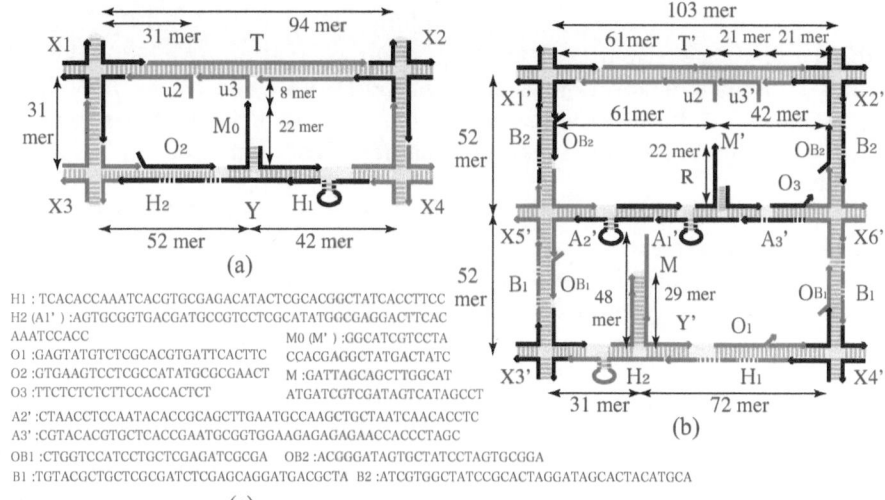

Fig. 6. The design of working-bases using 4-arm DNA junctions and 3-arm DNA junctions. (a) The working-basis consisting of a rectangle structure, (b) the working-basis consisting of two rectangle structures with a shared edge, (c) the sequences of some DNAs used in (a) and (b).

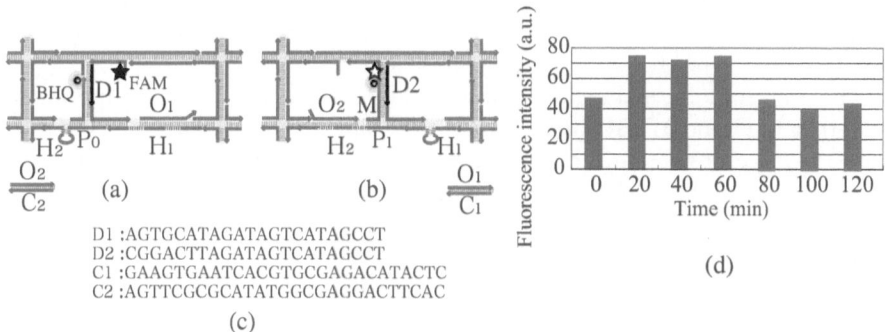

Fig. 7. Overview and result of the experiment on control of the position of M using conformation-change of a pair of hairpin-DNAs. The detection of the position using FRET when M is at (a) P_0 and (b) P_1. (c) The sequence of the DNA used for detect of the position and closing H_1 or H_2. (d) The fluorescence intensity measured with 20-minute intervals.

The components X1, X2, X3, X4, T, and Y were incubated by gradually changing temperature from 90 degrees C to 4 degrees C in the individual tubes. They were mixed into a tube and incubated at 4 degrees C to make the structure of the working-basis shown in Fig. 7 (b). At the beginning, strands O_1 and C_2

were added to the solution to move M from P_1 to P_0; and after 60 minutes, strands O_2 and C_1 were added to the solution to move M from P_0 to P_1. A sequence of FRET signal was detected at an interval of 20 minutes for 120 minutes. Just before detection, strands D_1 and D_2 were added for immobilizing the position of M. In the experiment, the solution were divided into seven tubes, and the different tubes were used to measure FRET signal at the respective detection times.

Figure 7 (d) shows measured fluorescence intensity on time. The difference of fluorescence indicates change of the distance between FAM and BHQ-1. Compared with the fluorescence at the beginning, fluorescence increases after 20 minutes. This shows movement of M from P_1 to P_0. The decrease of fluorescence at 80 minutes compared to that at 60 minutes indicates movement from P_0 to P_1. We confirmed that the position of M was controlled through the conformation-change of a pair of hairpin-DNAs. Using this operation, the positional information of M can be utilized to encode the state in the automaton. It can be considered that the reactions for moving M by adding strands O_1 and C_2 or adding strands O_2 and C_1 reach to a state of equilibrium within 20 minutes because fluorescence intensities at 20 through 60 minutes or at 80 through 120 minutes are similar. The time is closely related to the transition-rate of the automaton.

In the next experiment, we investigated the behavior of transmitting the positional information from the bottom tier to the top tier. The position of M and M' represent the current and the next state. The change of the position of M' is induced by binding of hairpin-DNAs with M and the following conformation-change of a pair of A_1' and A_3', or A_2' and A_3'. The information on the next state is transmitted to the top tier, and is hold there.

The procedure for making the working-basis is similar to the first experiment, but strands T' and M' are labeled with FAM and BHQ-1. Figures 8 (a) and (b) show the structure before and after the operation for transferring the information on the state to the middle tier when the initial state is S_0. This transferring starts by changing the open-state of B_1 to the closed-state. When B_1 is in the open-state, M cannot react with A_2', and no further reaction arises. As a result, M' is at the original position, and FRET occurs due to the proximity of two fluorophores as shown in Fig. 8(a). In this case, fluorescence of FAM is low. When B_1 is in the closed-state, M reacts to open A_2', and M' moves as shown in Fig. 8(b). Fluorescence of FAM is high because BHQ-1 is away from FAM. The change of conformations of A_3' and B_1 is induced by closer DNAs, C_3 and C_{B1}.

For fabricating the working-basis, the components {X1', X2', X3', X4', X5', X6', T', Y', R'} were mixed in a tube by the same procedure to the first experiment. The solution was divided into three tubes of solutions (i), (ii) and (iii). To obtain the baseline of fluorescence, solution (i) was prepared. The fluorescence of the solution (i) was measured just after mixing all the components not to make the working-basis. B_1 and B_2 are open-state at the initial. Closer DNA, C_{B2}, was added to solution (ii), and C_{B1} and C_{B2} were added to solution (iii). Strand E was added to all of the solutions for stabilizing the edge around A_1' or

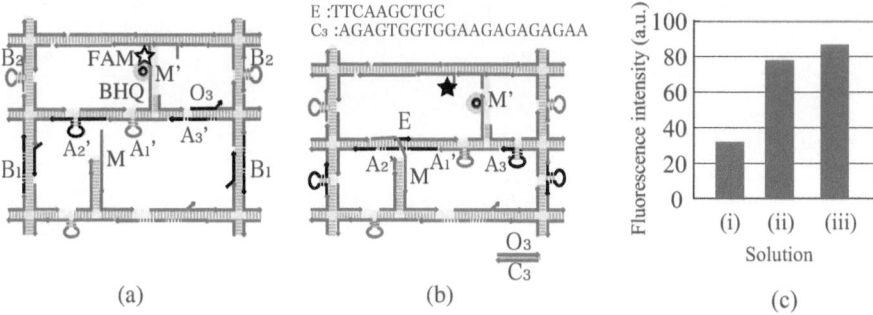

Fig. 8. The experimental scheme for transferring positional information of M. The DNA structures (a) before and (b) after transferring the positional information of M to the middle tier, (c) The fluorescence intensity obtained from solutions (i), (ii), and (iii).

A_2' witch was switched to the open-state. The expected structures of DNAs in solutions (ii) and (iii) are as shown in Figs. 8 (a) and (b), respectively.

Fluorescence intensities of solutions (i) through (iii) are shown in Fig. 8(c). Compared with fluorescence of solutions (i), fluorescence of solutions (ii) and (iii) is high. This result indicates that M reacts with hairpin-DNAs on the middle tier. It can be seen a little difference of fluorescence between solutions (ii) and (iii). This is expected to arise from the difference of distance between the bottom and the middle tiers. The information on the state can be transmitted from the bottom tier to the top tier, although the transmitting efficiency is small.

The fluorescent data obtained in the experiments support validity of some parts of our automaton. However, the data also suggest low control efficiency for the intended transition of the automaton. The possible reasons include unexpected behavior of the working structure and undesirable inter- and intra-molecular interactions. Direct observation of the structure using an AFM or other instrument will help to investigate the problems.

5   Conclusions

We demonstrated that information encoded as DNA conformation was able to be converted to positional information of DNA in a working-basis composed of DNA strands. Positional information of DNA is useful in representing the internal states of automaton to transmit information and to induce localized reactions in a DNA nano-structure. A function necessary for implementing a photonic DNA automaton was also confirmed using the working-basis. The internal state can be transmitted to the next tier in the working-basis by using the positional information. The result suggests that a portion of automata constructed by working-basis transit to the next state for a particular state-transition diagram and input. By using reaction schemes that produce different conformations of DNA depending on optical input, arbitrary state-transition diagrams with two kinds of symbols and two states will be implemented based on the photonic DNA automaton. Future issues include demonstration of state-transition using light and repetition of state-transitions.

Acknowledgments

This work was supported by the Ministry of Education, Culture, Sports, Science and Technology, Japan, a Grant-in-Aid for Scientific Research (A), 18200022, 2006-2008.

References

1. Benenson, Y., Gil, B., Ben-Dor, U., Adar, R., Shapiro, E.: An autonomous molecular computer for logical control of gene expression. Nature 429, 1–6 (2004)
2. Sakai, H., Ogura, Y., Tanida, J.: An Implementation of a Nanoscale Automaton Using DNA Conformation Controlled by Optics Signals. In: International Topical Meeting on Information Photonics 2008, pp. 172–173 (2008)
3. Sakai, H., Ogura, Y., Tanida, J.: An Implementation of a Nanoscale Automaton Using DNA Conformation Controlled by Optical Signals. Japanese Journal of Applied Physics (accepted, 2009)
4. Aldaye, F.A., Palmer, A.L., Sleiman, H.F.: Assembling Materials with DNA as the Guide. Science 321, 1794–1799 (2008)
5. Asanuma, H.: Synthesis of azobenzene-tethered DNA for reversible photo-regulation of DNA functions: hybridization and transcription. Nature Protocols 2, 203–213 (2007)
6. Kallenbach, N.R., Ma, R., Seeman, N.C.: An immobile nucleic acid junction constructed from oligonucleotides. Nature 305, 829–831 (1983)
7. Ma, R., Kallenbach, N.R., Sheardy, R.D., Petrillo, M.L., Seeman, N.C.: Three-arm acid junctions are flexible. Nucleic Acids Research 14, 9745–9753 (1986)
8. http://www.nupack.org/

Construction of AND Gate for RTRACS with the Capacity of Extension to NAND Gate

Yoko Sakai, Yoriko Mawatari, Kiyonari Yamasaki, Koh-ichiroh Shohda, and Akira Suyama

Department of Life Sciences and Institute of Physics
Graduate School of Arts and Sciences, The University of Tokyo
3-8-1 Komaba, Meguro-ku, Tokyo 153-8902, Japan
{sakai,mawatari,yamasaki,shohda}@genta.c.u-tokyo.ac.jp,
suyama@dna.c.u-tokyo.ac.jp

Abstract. We have succeeded in construction of the AND gate using enzymatic reactions developed for modularized computation elements of the autonomous computing system RTRACS. Experimental results demonstrated that the molecular reaction for the AND gate generated the correct output RNA from input RNAs according to the truth table for the AND gate. The constructed molecular reaction for the AND gate can be extended to the NAND gate by small modifications, because not only a logical 1 but also a logical 0 for inputs and output was associated with the presence of RNA strands.

Keywords: logic gate, molecular computer.

1 Introduction

RTRACS (Reverse-transcription and TRanscription-based Autonomous Computing System) is an autonomous DNA computing system proposed by learning from the replication mechanism of retrovirus [1,2]. RTRACS is composed of modularized computation elements that accept input RNA sequences associated with arguments and return output RNA sequences associated with return values. Conversion of input RNA sequences to output RNA sequences are performed at a constant temperature through building-reactions including hybridization, reverse transcription, RNA digestion, DNA polymerization, and transcription. RTRACS can be built up to meet demanding problems by combining modularized computation elements that can execute basic functions such as AND, OR, NOT, etc. The network of the computation elements connected by their input and output RNA sequences can form a computational circuit similar to integrated circuits in electronic computers (Fig. 1).

In the previous work we reported the construction of the molecular reaction for AND gates using a logical 0 of inputs and output associated with the absence of the RNA strands [2]. This implementation, however, does not allow us to construct modularized computation elements executing the NOT operation. Therefore, no NAND gate can be constructed by extending the previous AND

R. Deaton and A. Suyama (Eds.): DNA 15, LNCS 5877, pp. 137–143, 2009.

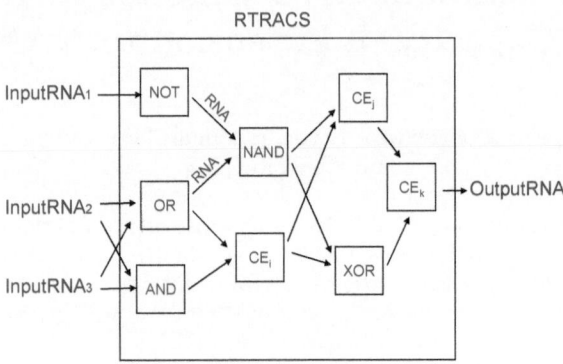

Fig. 1. RTRACS (Reverse-transcription and TRanscription-based Autonomous Computing System). RTRACS is a network of modularized computation elements (designated by boxes □) connected with their input and output RNA strands (designated by linear arrows ⟶). The input RNA sequences of a computation element contain the values and addresses of inputs, and the output RNA sequences have those of outputs. The connection between modularized computation elements is built up by using the addresses of their inputs and outputs. Computation on RTRACS is executed on a network of modularized computation elements. Therefore, the structure of the network represents a program of RTRACS.

gate. In this study, we have constructed the molecular reaction for the AND gate in which not only a logical 1 but also a logical 0 of inputs and output was associated with the presence of RNA strands. This implementation makes substantial improvements on RTRACS, because it allows the NOT operation to be executed by just converting a RNA sequence associated with a logical 1 (or 0) into one with a logical 0 (or 1). Therefore, the present molecular reaction for the AND gate has the capacity for extension to the NAND gate.

2   Reaction Scheme of AND Gate

The AND gate is the all or nothing gate, the operation of which is defined by the truth table in Table 1. The truth table for the AND gate gives all the possible input combinations of X and Y and the resulting outputs of Z. The table says that only when both inputs X and Y are 1 output Z is 1 while the rest of the input combinations give outputs $Z = 0$. In electronic AND gates, a logical 0 or logical 1 is determined using input and output voltage levels. In the present AND gate for RTRACS, a logical 0 or logical 1 is determined using input and output RNA sequences.

Figure 2 indicates the reaction scheme of the AND gate constructed in this study. RNA sequences associated with inputs $X = 1$, $X = 0$, $Y = 1$, and $Y = 0$ are y-$1x$, $0z$-$0x$, $1z$-y, and $0z$-y, respectively. Symbols y, $1x$, $0x$, $1z$, and $0z$ stand for sequences chosen from the set of orthonormal sequences developed for reliable

Table 1. Truth table for AND gate

Inputs		Output
X	Y	Z
1	1	1
1	0	0
0	1	0
0	0	0

DNA computation. RNA sequences associated with $Z = 1$ and $Z = 0$ depend on the connection of the output with the input of a subsequent AND gate. When the Z output is connected with the Y input of a subsequent gate designated by Y', output RNA *1z'-y'* and *0z'-y'* are associated with $Z = 1$ and $Z = 0$, respectively.

The molecular reaction for the AND gate starts with hybridization of DNA primer *c1x* or *c0x* with input RNA associated with X. As shown in Fig. 2a, if inputs are $X = 1$ and $Y = 1$, primer *c1x* hybridizes with input RNA *y-1x* ($X = 1$) and initiates a DNA strand synthesis with a reverse transcriptase using the input RNA as a template. The RNA strand of the resultant DNA/RNA hetero-duplex is digested with RNase H, and single-stranded DNA (ssDNA) *c1x-cy* of 52 bases long is generated. The 52 mer ssDNA then hybridizes with input RNA *1z-y* ($Y = 1$) and initiates a reverse transcription of DNA strand using the input RNA as a template. The 78 mer ssDNA *c1x-cy-c1z* consequently produced has a sequence *c1z* at the 3'-end, which indicates the result of ANDing input $X = 1$ and $Y = 1$. The sequence *c1z* hybridizes with a converter DNA that produces output RNA associated with $Z = 1$. A reverse-transcriptase, which can synthesize DNA not only using a RNA template but also using a DNA template, performs a primer extension to activate the converter DNA by making a T7 promoter region (designated by *T7P*) double-stranded. Output RNA *1z'-y'* associated with $Z = 1$ is finally produced by transcription with T7 RNA polymerase.

The molecular reaction for the AND gate proceeds in a similar manner if inputs are $X = 1$ and $Y = 0$ (Fig. 2b). Primer *c1x* starts a series of reactions for ANDing inputs X and Y, and a 52 mer ssDNA *c1x-cy* is generated by reverse-transcription of input RNA *y-1x* ($X = 1$). The resultant 52 mer is then extended to a 78 mer ssDNA *c1x-cy-c0z* with input RNA *0z-y* ($Y = 0$). At the 3'-end, ssDNA *c1x-cy-c0z* has a sequence *c0z* that indicates the result of ANDing input $X = 1$ and $Y = 0$. The sequence *c0z* hybridizes with a converter DNA that produces output RNA associated with $Z = 0$. The converter DNA activated by primer extension with *c0z* finally generates transcripts of output RNA *0z'-y'* associated with $Z = 0$.

The molecular reaction for the AND gate proceeds in a slightly different manner if inputs are $X = 0$ (Fig. 2c). In this case, Z is always 0 and does not depend on Y inputs. Therefore, RNA strands for the Y inputs are not involved in the molecular reaction at all. Primer *c0x* starts a series of reactions for ANDing inputs X and Y, and a 52 mer ssDNA *c0x-c0z* is generated by reverse-transcription of input RNA *0z-0x* ($X = 0$). The resultant 52 mer ssDNA has a sequence *c0z*

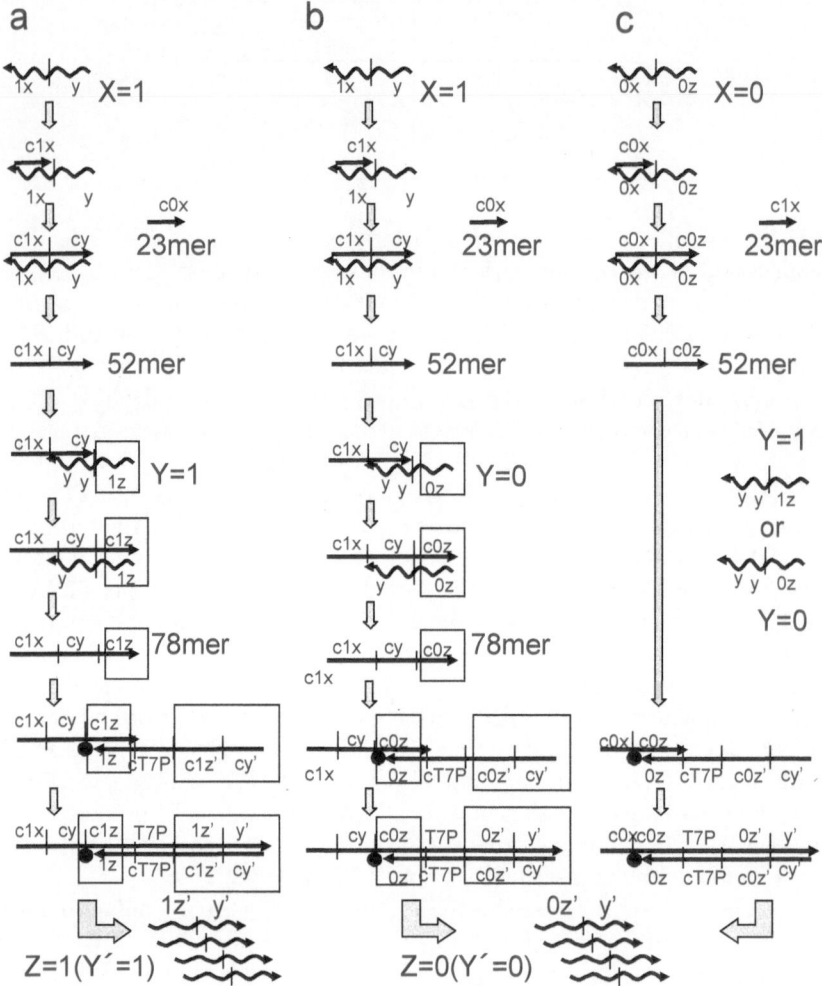

Fig. 2. Reaction scheme of AND gate. (a) Inputs $X = 1$ and $Y = 1$. (b) Inputs $X = 1$ and $Y = 0$. Solid line boxes indicate the difference between $(X = 1, Y = 1)$ and $(X = 1, Y = 0)$. (c) $X = 0$ and $Y = 1$ or 0. Linear arrows $\longrightarrow$ and wavy arrows $\rightsquigarrow$ indicate DNA and RNA strands, respectively. Solid dots • designate NH_2, which is attached to the 3'-end of DNA strands to avoid unwanted strand extension with reverse transcriptase.

indicating the result of ANDing input $X = 0$ and $Y = 1/0$ at the 3'-end. The sequence $c0z$ hybridizes with a converter DNA that produces output RNA associated with $Z = 0$. The activated converter DNA finally generates transcripts of output RNA $0z'$-y'.

3   Results

The molecular reaction for the AND gate has been investigated to confirm that
the reaction proceeded as indicated by the scheme in Fig. 2. The reaction can
be divided into two parts. The first part is a series of reactions started with
the hybridization of primer DNA $c1x$ or $c0x$ with input X and continued until
the creation of ssDNA with a sequence indicating the result of ANDing inputs
X and Y at the 3'-end. The remaining part is the reactions that activate a
converter DNA and generate output RNA transcripts. We first examined the
first part of the reaction using reaction mixtures without the converter DNA,
and then examined the input-output relation of the AND gate reaction using
reaction mixtures containing all components.

The first part of the AND gate reaction was examined using fluorescence
dye-labeled DNA primers, Cy5-labeled $c1x$ and Cy3-labeled $c0x$. Cy5 and Cy3
have different fluorescence colors. Cy5 emits red fluorescence and Cy3 emits
green fluorescence. Thus DNA strands extended with these primers can be easily
distinguished from each other.

The reaction scheme in Fig. 2 indicates that if X inputs are 1 primer DNA
$c1x$ is extended to a 52 mer ssDNA and then to a 78 mer ssDNA regardless
of Y inputs. Primer DNA $c0x$ is not involved in the reaction and results in no
extension. If X inputs are 0, instead of $c1x$, primer DNA $c0x$ is involved in the
reaction and generates a 52 mer ssDNA. This ssDNA is not subjected to a further

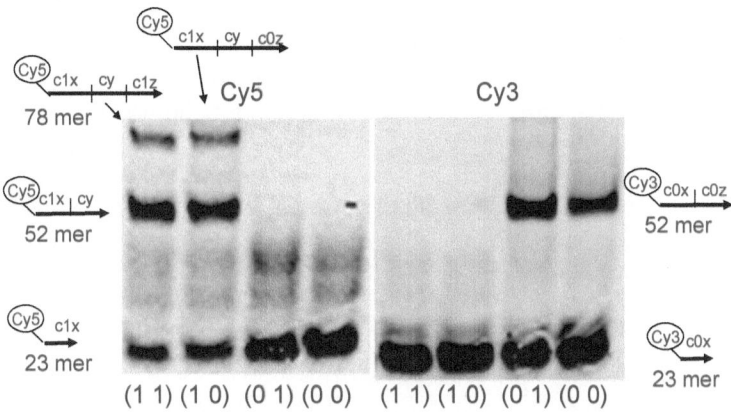

Fig. 3. Denaturing gel electrophoresis of resultant ssDNA labeled with fluorescence
dye Cy5 or Cy3 after a 2 h reaction at 50 °C. Reaction mixtures with RNA inputs
$X = 1$ and $Y = 1$, $X = 1$ and $Y = 0$, $X = 0$ and $Y = 1$, $X = 0$ and $Y = 0$ are run
on lanes (1 1), (1 0), (0 1), and (0 0), respectively. The electrophoresis was carried out
on a denaturing gel containing 12% polyacrylamide (acrylamide/ bisacrylamide, 29:1)
and 8.0 M urea in TBE electrophoresis buffer at 65 °C. After electrophoresis Cy5 and
Cy3 images were obtained from the same gel using a fluoroimage analyzer, FLA-5100
(Fuji Film, Japan).

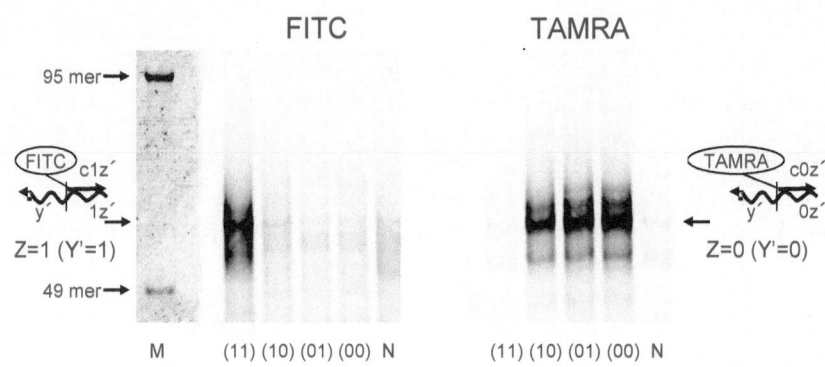

Fig. 4. Native gel electrophoresis of output RNA transcripts hybridized with fluorescence dye-labeled probes. The AND gate reaction was continued for 2 h at 50 °C. FITC-labeled probe and TAMRA-labeled probe specifically hybridize with output RNA transcripts associated with $Z = 1$ and $Z = 0$, respectively. Reaction mixtures with RNA inputs $X = 1$ and $Y = 1$, $X = 1$ and $Y = 0$, $X = 0$ and $Y = 1$, $X = 0$ and $Y = 0$ are run on lanes (1 1), (1 0), (0 1), and (0 0), respectively. N lane is a reaction mixture without input RNA. M lane is marker DNA. The electrophoresis was carried out on a gel containing 12% polyacrylamide (acrylamide/ bisacrylamide, 29:1) in TBE electrophoresis buffer. After electrophoresis gel images were obtained using a fluoroimage analyzer, FLA-5100.

extension, so that only the 52 mer ssDNA is generated in a reaction mixture. The lengths of Cy5- and Cy3-labled ssDNA were analyzed by denaturing-gel electrophoresis. The result consistent with the reaction scheme was obtained as shown in Fig. 3. Therefore, the molecular reaction for the AND gate proceeded in a manner as indicated in Fig. 2.

The input-output relation of the AND gate reaction was examined by electrophoresis of output transcripts. The sequences of the transcripts were determined using fluorescence dye-labeled probes specific to output RNA sequences. A FITC-labeled $c1z'$ probe, which has a green fluorescent color, can specifically hybridize with output RNA $1z'\text{-}y'$ associated with $Z = 1$. To specifically detect output RNA $0z'\text{-}y'$ associated with $Z = 0$, a TAMRA-labeled $c0z'$ probe with a red fluorescent color was used.

Figure 4 shows the result of experiments to examine the input-output relation of the AND gate reaction. If inputs are $X = 1$ and $Y = 1$, output RNA $1z'\text{-}y'$ with a length of 52 bases was detected. For the rest of input combinations, namely, $X = 1$ and $Y = 0$, $X = 0$ and $Y = 1$, $X = 0$ and $Y = 0$, output RNA $0z'\text{-}y'$ with a length of 52 bases was detected instead of $1z'\text{-}y'$. Output RNA $1z'\text{-}y'$ and $0z'\text{-}y'$ are associated with $Z = 1$ and $Z = 0$, respectively. Therefore, the input-output relation of the AND gate reaction was proved to be consistent with the truth table for the AND gate in Table 1.

4   Discussion

We have succeeded in construction of the AND gate using enzymatic reactions developed for modularized computation elements of the autonomous computing system RTRACS. Experimental results in Figs. 3 and 4 demonstrate that the molecular reaction for the AND gate designed as shown in Fig. 2 generated the correct output RNA from input RNAs according to the truth table for the AND gate. The constructed AND gate could work as a modularized computation element in RTRACS, because output RNA has a sequence acceptable to another AND gate as input RNA.

In the previous construction of the AND gate reaction, the absence of RNA strands is associated with a logical 0 for inputs X and Y and output Z [2]. In contrast, the present construction requires the presence of RNA strands associated with a logical 0 for the inputs and output. This requirement makes substantial improvements on RTRACS, because it allows the NOT operation to be executed. By converting a RNA sequence associated with a logical 1 (or 0) into one with a logical 0 (or 1), the NOT operation can be performed. Therefore, the molecular reaction for the NAND gate, which is a universal logic gate, can be constructed from the present AND gate reaction.

The present construction of the AND gate also offers substantial advantages in application of RTRACS to genetic diagnosis. If the absence of RNA is associated with a logical 0, there is no other way but to determine the result of a diagnosis by the amount of a single output RNA transcript associated with $Z = 1$. This diagnosis method is not robust, because many factors, including enzyme activities and sample amounts, affect the output RNA amount. Determination of a diagnosis result by comparing the amounts of two output RNA transcripts associated with $Z = 1$ and $Z = 0$ is more robust, because the ratio of the two amounts is less affected by enzyme activities and sample amounts. A single-nucleotide polymorphism (SNP) pattern diagnosis method based on the present AND gate reaction is under development.

5   Conclusion

We have succeeded in construction of an AND-gate molecular reaction that can be extended to the NAND gate.

Acknowledgments. This work was supported by a grant for SENTAN (Development of Systems and Technology for Advanced Measurement and Analysis) from the Japan Science and Technology Agency (JST), and by a grant-in-aid from the Ministry of Education, Culture, Sports, Science, and Technology of Japan.

References

1. Nitta, N., Suyama, A.: Autonomous Biomolecular Computer Modeled after Retroviral Replication. In: Chen, J., Reif, J.H. (eds.) DNA 2003. LNCS, vol. 2943, pp. 203–212. Springer, Heidelberg (2004)
2. Takinoue, M., Kiga, D., Shohda, K., Suyama, A.: Experiments and Simulation Models of a Basic Computation Element of an Autonomous Molecular Computing System. Phys. Rev. E 78, 041921 (2008)

Time-Complexity of Multilayered
DNA Strand Displacement Circuits

Georg Seelig[1] and David Soloveichik[2]

[1] University of Washington, Seattle, WA, USA
[2] California Institute of Technology, Pasadena, CA, USA
gseelig@u.washington.edu,dsolov@caltech.edu

Abstract. Recently we have shown how molecular logic circuits with many components arranged in multiple layers can be built using DNA strand displacement reactions. The potential applications of this and similar technologies inspire the study of the computation time of multilayered molecular circuits. Using mass action kinetics to model DNA strand displacement-based circuits, we discuss how computation time scales with the number of layers. We show that depending on circuit architecture, the time-complexity does not necessarily scale linearly with the depth as is assumed in the usual study of circuit complexity. We compare circuits with catalytic and non-catalytic components, showing that catalysis fundamentally alters asymptotic time-complexity. Our results rely on simple asymptotic arguments that should be applicable to a wide class of chemical circuits. These results may help to improve circuit performance and may be useful for the construction of faster, larger and more reliable molecular circuitry.

Circuit depth is the standard measure of time-complexity of feed-forward circuits [8]. While this is well justified in electronic digital circuits, in this paper we ask whether depth is the correct measure of time-complexity for chemical circuits. We provide a quantitative analysis of how computation time is related to circuit size and architecture. We compare two elementary mechanisms for the underlying components: in one case, the underlying chemical reactions are stoichiometric and one input molecule produces one output molecule. In the other case the underlying reactions are catalytic and a single input molecule can trigger an arbitrary number of output molecules. We show that for non-catalytic circuits, the time to half-completion does not always scale linearly with the depth of the circuit. Our analysis shows that for a tree of stoichiometric bimolecular reactions, the time to half-completion scales quadratically with the depth of the circuit — i.e. the additional time due to adding an extra layer increases linearly with the size of the circuit. In contrast, we find that for catalytic systems the time to half-completion is a linear function of the depth independently of the structure of the circuit. The latter results agrees with our intuition from electronics where all gates are amplifying.

In this paper, for the physical model of molecular circuits we focus on DNA-based circuits implemented as cascades of strand displacement reactions.

R. Deaton and A. Suyama (Eds.): DNA 15, LNCS 5877, pp. 144–153, 2009.
© Springer-Verlag Berlin Heidelberg 2009

Single-stranded nucleic acids serve as signals that are exchanged between multi-stranded gate complexes. We have previously shown that this technology allows us to build multi-component molecular circuits that incorporate all the main features of digital logic, such as Boolean logic gates like AND, NOT and OR, signal restoration and modularity [4]. More recent papers have implemented a variety of improvements including gates for fast catalytic amplification [11,9], and reversible logic gates based on a simple catalytic gate motif [3]. This technology provides a starting point for building large-scale molecular circuitry with quantitatively predictable behavior using standardized off-the-shelf components [6].

In our experiments on nucleic-acid logic circuits we noticed that often the measured time to half-completion did not seem to scale linearly with the depth of the circuit. Instead, every additional layer seemed to add more than a constant offset to the time to half-completion. The observed slowdown may in part be due to non-specific interactions between DNA species that compete with or hinder the desired interactions between DNA gates and their inputs. However, here we will argue that the observed slowdown is at least in part a consequence of the circuit layout used in these experiments and of the underlying reaction kinetics of the DNA components.

In the next section we compare the time-complexity of linear cascades and converging trees for both non-catalytic and catalytic reactions. These circuit architecture represent extreme cases: A linear cascade of length N minimizes total fan-in while the converging tree of equal depth maximizes total fan-in. We describe specific implementations of these circuits with previously developed components and use numerical simulations to investigate how time complexity relates to circuit architecture and reaction mechanism. Then, we derive the asymptotic scaling of time-complexity of circuits using simplified kinetics. The simplified kinetics captures the essential features of the strand displacement circuits, but should be also applicable to alternative implementations of molecular circuits. Indeed the proofs are largely independent of the details of the underlying reaction mechanisms.

1   Comparing DNA Strand-Displacement Base Reaction Mechanisms and Circuit Architectures

In this section we consider linear reaction cascades and converging trees based on DNA strand displacement chemistry previously described [4]. In the next section we consider more abstract and more generally applicable models of the same architectures that capture the essential behavior.

1.1   Cascades of Non-catalytic Reactions

A strand displacement reaction (see Fig. 1(A)) can be intuitively described as a hybridization reaction between two complementary strands where one strand is initially fully single stranded (the "input") while the other strand is already partially double stranded (the "gate"). The reaction is driven forward by the

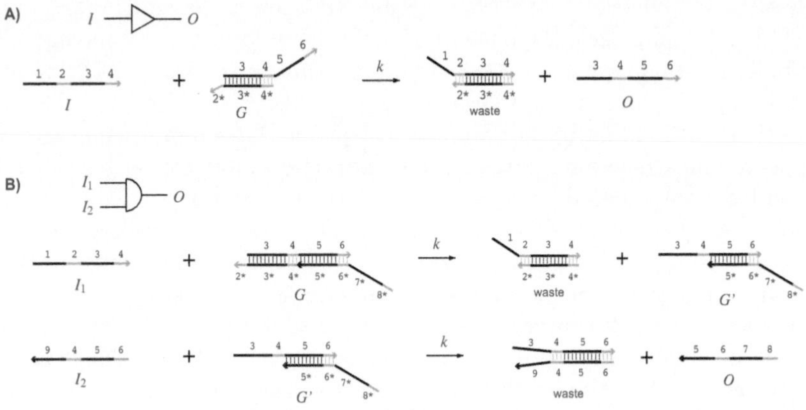

Fig. 1. (A) Strand displacement transducer gate, producing signal O if signal I is present. (B) Strand displacement AND gate, producing signal O only if I_1 and I_2 are present. Toehold domains are shown in red. AND gates with different 3'-5' input and output orientations can be constructed similarly.

formation of extra base pairs in the final compared to the initial state. Even though the underlying reaction mechanism is more complex, the overall reaction kinetics is well described as a bimolecular reaction $I + G \xrightarrow{k} O$ [10]. The rate constant k can be adjusted by changing the length and sequence composition of the single-stranded toe-hold (i.e. the single stranded overhang on the gate complex). If both reactants are present at the same initial concentration, i.e. $I(0) = G(0) = g_0$, the time evolution of G is given by the well-known expression

$$G(t) = \frac{g_0}{1 + kg_0 t} \tag{1}$$

The time to half-completion $\tau_{1/2}$ for this reaction is $\tau_{1/2} = 1/kg_0$.

We now consider a linear cascade of strand displacement reactions of the type shown in Fig. 1. Fig. 2(A) shows the corresponding circuit diagram. For the n-th layer in the cascade we have the reaction $I_n + G_n \xrightarrow{k} I_{n+1}$. We have already solved the case $n = 1$ and for $n > 1$ conservation of mass requires that $I_n = G_n - G_{n-1}$. The resulting equation of motion is

$$\dot{G}_n = -k \left(G_n - G_{n-1} \right) G_n. \tag{2}$$

This equation has the form of a Bernoulli differential equation and can therefore be integrated formally. If we assume that all gates G_n and the initial input I_1 are present at the same initial concentration $I_1(0) = G_n(0) = g_0$ we can obtain a closed form solution for arbitrary n. In this case the result becomes

$$G_n(t) = g_0 \left[1 - \frac{\frac{1}{n!} (kg_0 t)^n}{\sum_{m=0}^{n} \frac{1}{m!} (kg_0 t)^m} \right]. \tag{3}$$

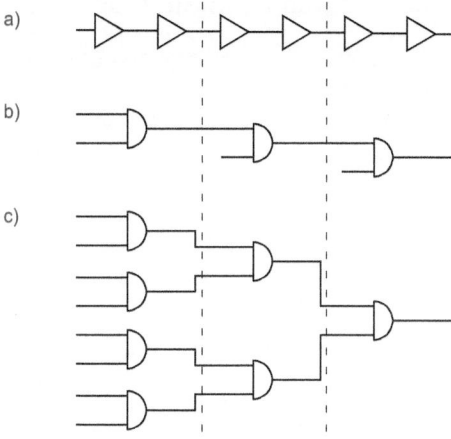

Fig. 2. Circuit architectures: (A) Cascade of tranducer gates. (B) Linear cascade of AND gates. (C) Maximally converging tree of AND gates. The alignment indicates that activation of an AND gate requires two sequential strand displacement reactions while a activation of a transducer gate requires one reaction.

We can now ask at what time τ_f^N a particular fraction fg_0 ($0 < f < 1$) of the maximum output level has been reached at the final layer N. The output $I_N(t)$ of the last reaction in the cascade is $I_N(t) = g_0 - G_N(t)$, and the time τ_f^N is the solution to $I_N(\tau_f^N) = fg_0$. Solving this equation for τ_f^N exactly would require us to find the zeroes of a polynomial of order N which is generally not possible. However, it can be shown[1] that $\tau_f^N = \frac{1}{1-f}\frac{N}{kg_0}$ is a good solution for large N in the sense that $\lim_{N\to\infty} I_N(\frac{1}{1-f}\frac{N}{kg_0}) = fg_0$. Then the cost in terms of additional run-time of adding an extra layer to a cascade of length N, for large N, can be well approximated as $\tau_f^{N+1} - \tau_f^N = \frac{1}{1-f}\frac{1}{kg_0}$ which means that adding an additional layer only leads to a constant (independent of N) increase in computation time. This agrees with the results from our numerical simulations (see Fig. 3).

---

[1] The proof technically reduces to showing

$$\lim_{N\to\infty} \frac{(\alpha N)^N}{N!} \Big/ \sum_{m=0}^{N} \frac{(\alpha m)^m}{m!} = \frac{\alpha - 1}{\alpha} \tag{4}$$

for $\alpha \geq 1$. We can prove that the limit is at least $\frac{\alpha-1}{\alpha}$ by substituting $N^N/N! > m^m/m!$ in the denominator. We upper-bound the limit by discarding some initial terms in the denominator. Specifically, for any ϵ we can choose a large enough k that $\alpha^k / \sum_{m=0}^{k} \alpha^m$ is within $\epsilon/2$ of $\frac{\alpha-1}{\alpha}$, and for large enough N, the left hand side of eq. 4, keeping only the last $k+1$ terms, is within $\epsilon/2$ of $\alpha^k / \sum_{m=0}^{k} \alpha^m$.

1.2   Converging Trees of Non-catalytic Reactions

When two strand displacement reactions are cascaded on the same DNA complex, we obtain the second most simple gate motif: the AND gate of Ref. [4] shown in Fig. 3(B). The two inputs to the AND gate react sequentially and in an order specified by the gate design. The strand displacement reactions between the AND gate its two inputs are

$$I_1 + G \xrightarrow{k} G', \tag{5}$$

$$I_2 + G' \xrightarrow{k} O. \tag{6}$$

A linear chain of N AND gates where the output of a gate in layer n is the first inputs to the gate in layer $n+1$ while the second input to each gate is provided initially is an alternative implementation of a strand-displacement reaction cascade of depth $2N$ (see Fig. 2 (B)). It can be analyzed using the same reaction equations, but with the roles of activated gate complex G' and single stranded input switched in the even layers.

As a contrasting architecture to a linear cascade, we consider a converging tree of AND gates that is N gates deep and $2N$ reactions deep as indicated in Fig. 2 (C). In this case, both inputs to an AND gate in any layer $n > 1$ must be the output of an AND gate in the previous layer. While the linear cascade has uniform fan-in 1, this circuit has uniform fan-in 2.

Simulation results for converging trees with up to 50 layers are shown in Fig. 3 (B). Interestingly, and unlike in the previous examples the time-complexity is not a linear function of the number of layers but in fact appears to increase quadratically with the depth of the circuit.

1.3   Cascades of Catalytic Reactions

Cascades of catalytic reactions occur in many biological networks such as metabolic or signaling pathways. A variety of rationally designed DNA catalytic gates and even cascades of catalytic gates have been experimentally implemented [7,5,11,2,1]. In the following we use the minimal model $I_1 + G_1 \xrightarrow{k} I_1 + I_2$ to describe the catalytic systems. The reaction mechanisms proposed for hybridization-based catalytic systems in the literature [7,5,11] typically involve multiple strand displacement steps, but this simple model should capture the key differences between catalytic and non-catalytic strand displacement.

In contrast to the non-catalytic case, each step in a catalytic reaction cascade is intrinsically amplifying. A single input or catalyst can react sequentially with multiple gates or substrates and one would therefore expect that such cascades transmit signals faster. Similarly, since a sub-stoichiometric amount of input is enough to catalytically turn over a large number of substrates, we would expect the approach to equilibrium to be exponentially fast. A numerical simulation of a cascade of catalytic reactions confirms this intuition. The simulation provides strong evidence that the time-complexity grows linearly with the number of layers (see Fig. 3).

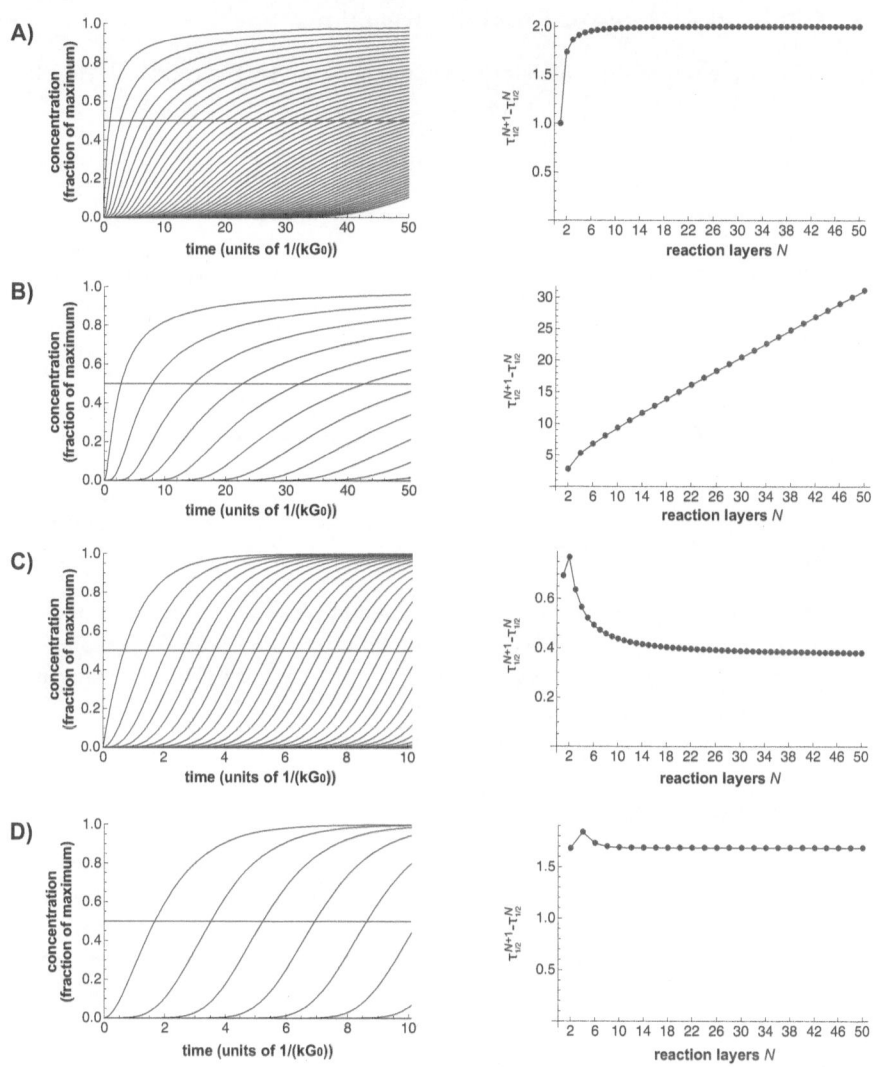

Fig. 3. Numerical simulations of different network topologies: (A) linear cascade, (B) converging exponential tree, (C) catalytic linear cascade, (D) catalytic converging exponential tree. For each case we simulated circuits of reaction depth $N = 1$–50. Note that in the converging exponential tree, a single circuit layer corresponds to two reactions. (Left column) Time-evolution of the output O_N for all circuits ($N = 1$–50). Note the vastly different timescales of (A)-(D). (Right column) $\tau_{1/2}^{N+1} - \tau_{1/2}^{N}$ as a function of N. Note that unlike in the case of a linear cascade (A) or catalytic systems (C)-(D), adding each new layer to the converging exponential tree without catalysis increases the half-time to completion by a larger and larger interval.

1.4   Converging Trees of Catalytic Reactions

Finally, we want to consider a converging tree of catalytic reactions. We consider
a tree using an ordered AND gate where each input acts catalytically, i.e.

$$I_1 + G \xrightarrow{k} I_1 + G', \tag{7}$$

$$I_2 + G' \xrightarrow{k} I_2 + O. \tag{8}$$

As in the case of a non-catalytic AND gate, G' is a partially double-stranded
reaction intermediate. A converging tree of catalytic AND gates can be based on
the same composition rules we used for the non-catalytic case (see Fig. 2(C)).
However, unlike for the non-catalytic case, the time-complexity of a converging
tree is linear in the depth of the tree as shown by numerical simulations (see
Fig. 3). The systematic use of catalysis can therefore lead to a considerable
speed-up of signal propagation.

2   Asymptotic Arguments

In this section we derive the asymptotic time complexity of linear cascades and
converging trees, using simplified reaction kinetics. The simplified kinetics is
sufficient to explain the scaling of the time complexity with the increasing number of layers and desired completion fraction of the strand displacement circuits
considered above, but should be more generally applicable.

2.1   Non-catalytic Reactions

For the non-catalytic case, we only consider a converging reaction tree. (As
shown in Sec. 2.1 the differential equations describing a linear cascade can be
solved exactly and the asymptotic time complexity is linear in the depth of the
cascade.) As the simplest and most general non-catalytic AND gates we use
the reaction schema $I_1 + I_2 \to O$. We prove that for maximally converging trees
composed of these gates the time to f-fraction completion scales inversely as a
function of $1 - f$ and quadratically as a function of N, establishing both lower
and upper bounds. For simplicity we assume that the initial concentration of
inputs at the first layer is 1 and all rate constants are 1.

Lemma 1. *Fixing the number of layers, the time to f-fraction completion is
$\Omega(1/(1-f))$ as $f \to 1$. Fixing $0 < f < 1$, the time to f-fraction completion is
$\Omega(N^2)$ as $N \to \infty$.*

Proof. To achieve f-fraction completion, the f-fraction must pass through at
least the first layer, which takes $f/(1-f) = \Omega(1/(1-f))$ time, proving the first
part. For the second part, the time to f-fraction completion can only decrease if
we start in the state where each signal species in every layer is at $1/N$ concentration. Now note that in order for the amount of signal species in a given layer
to increase, the amount of each signal species in the previous layer must exceed

the amount of each signal species in the given layer. Thus starting in our start state, no signal species can exceed $1/N$ concentration. Then, looking at the last layer, the rate of the production of the final output is never more than $1/N^2$ and it takes at least $\Omega(N^2)$ time to reach f-fraction completion.

Lemma 2. *The time to f-fraction completion for N layers is no more than $N^2/(1-f)$.*

Proof. Consider the process in which we wait, sequentially, for each layer n to convert $1-(1-f)n/N$ amount of each input signal species to each output species. This process reaches f-fraction completion at the final layer no faster than the original process. Now the time for a layer starting with amount a of each input species to produce amount $b < a$ of each output species is $(b/a)/(a-b) < 1/(a-b)$. Thus, in our new process each layer takes less than $N/(1-f)$ time to convert the desired amount of input to output, for a total f-fraction completion time of the whole circuit of less than $N^2/(1-f)$.

2.2   Catalytic Reactions

Next we consider our two examples of catalytic circuits: catalytic linear chains and catalytic converging trees using the simplest models conserving mass. The linear catalytic chain is composed of reactions $I + G \rightarrow I + O$ where I is the input, O is the output, and G is the source of mass for the output. The catalytic converging tree is composed of AND modules consisting of reaction $I_1 + I_2 + G \rightarrow I_1 + I_2 + O$, where I_1 and I_2 are the two inputs, O is the output, and G is the source of mass. We assume all G start at 1 so that the maximum output level is equal to the input level. Note that the following results can be easily applied to different implementations since they do not rely on the details of the reaction mechanism.

We now show that the time for f-fraction completion of the catalytic linear chain and the catalytic converging tree scales linearly with the number of layers N and logarithmically in $1/(1-f)$.

Lemma 3. *The time to f-fraction completion is $O(N+\log(1/(1-f)))$ as $f \rightarrow 1$.*

Proof. In catalytic circuits, the amount of input for any layer can never decrease. First of all, the time for a layer, assuming fixed amount a of each input species, to produce amount b of each output species, is $\log(1/(1-b))/a$ for the linear catalytic chain, and $\log(1/(1-b))/a^2$ for the tree. Now consider the process in which we wait for the previous layer to produce $1/2$ of each output species, stop it, and start the next layer; this process is slower than the original process. In the new process, the total $1/2$-fraction completion time for the final layer is $O(N)$. Then in extra $O(\log(1/(1-f)))$ time f-fraction completion can be reached.

Lemma 4. *Fixing $0 < f < 1$, the time to f-fraction completion is $\Omega(N)$ as $N \rightarrow \infty$.*

Proof. Consider the catalytic linear chain since it is faster than the catalytic tree. Now consider the process in which the G species are never used up and always stay at their initial value 1. Clearly, the f-fraction completion time for this process is faster than that for the original. Now for the modified process the concentration of the output of layer $n-1$ at time t is the derivative of the concentration of the output of layer n at time t. Thus the concentration of output at layer n at time t is $t^n/n!$. By Sterling's formula, $\lim_{n\to\infty}(n/e)^n/n! = 0$, and thus we would not be able to reach the desired f-fraction for some sufficiently large number of layers N in time N/e.

3   Discussion

Motivated by the recent advances in DNA nanotechnology which allowed for the construction of complex biochemical circuits, we investigated the time-complexity of multilayered molecular circuits. We considered two different circuit architectures, namely linear reaction cascades and converging trees of chemical reactions of increasing depth. In each case, we compared catalytic and non-catalytic mechanisms for the underlying reactions.

First, we investigated time complexity through numerical simulations of multilayered circuits. We used a model of circuit components based on cascaded strand displacement reactions. To obtain analytic confirmation of the observed behavior, we derived asymptotic scaling relationships for a simplified kinetics model. The simplified model captures the essential behavior of the strand displacement cascades, but is general enough to apply to a wide class of circuits. Simulation and analytic results show that the time to (f-fraction) completion scales linearly with the number of layers for all considered architectures except non-catalytic converging trees. In the latter case, the corresponding completion time scales quadratically with the number of layers.

The approach to completion is increasingly slow with increasing depth of the circuit but the asymptotic approach to equilibrium behaves differently for catalytic and non-catalytic circuits: The time to f-fraction completion scales as $1/(1-f)$ for non-catalytic systems considered, but as $\log(1/(1-f))$ when catalysis is introduced.

From these observation we can conclude that the systematic use of catalysis allows one to build faster circuits with considerably improved asymptotic behaviors. However, care has to be taken since in a catalytic circuit any small leak at the input level will get exponentially amplified leading to a false positive output. A functional catalytic circuit will therefore require some degree of signal restoration which for example can be achieved through competitive inhibition (note that the additional components necessary for signal restoration can potentially slow down circuit operation). In contrast since there is no amplification in a cascade of bimolecular reactions the level of the output will not exceed the level of input.

Acknowledgments. We thank Ho-Lin Chen, Matthew Cook, Anton Andreev, Bernard Yurke and Erik Winfree for discussions and help. GS was supported by the Swiss National Science Foundation and by a Career Award at the Scientific Interface from the Burroughs Wellcome Fund.

References

1. Bois, J., Venkataraman, S., Choi, H., Spakowitz, A., Wang, Z., Pierce, N.: Topological constraints in nucleic acid hybridization kinetics. Nucleic Acids Research 33(13), 4090 (2005)
2. Green, S., Lubrich, D., Turberfield, A.: DNA hairpins: fuel for autonomous DNA devices. Biophysical Journal 91(8), 2966–2975 (2006)
3. Qian, L., Winfree, E.: A simple DNA gate motif for synthesizing large-scale circuits. In: Proceedings of the 14th International Conference on DNA Computing (2008)
4. Seelig, G., Soloveichik, D., Zhang, D., Winfree, E.: Enzyme-free nucleic acid logic circuits (2006)
5. Seelig, G., Yurke, B., Winfree, E.: Catalyzed relaxation of a metastable DNA fuel. Journal of the American Chemical Society 128(37), 12211–12220 (2006)
6. Soloveichik, D., Seelig, G., Winfree, E.: DNA as a Universal Substrate for Chemical Kinetics (Extended Abstract). In: Proceedings of the 14th International Conference on DNA Computing (2008)
7. Turberfield, A., Mitchell, J., Yurke, B., Mills Jr., A., Blakey, M., Simmel, F.: DNA Fuel for Free-Running Nanomachines. Physical Review Letters 90(11), 118102 (2003)
8. Vollmer, H.: Introduction to Circuit Complexity: A Uniform Approach. Springer, Heidelberg (1999)
9. Yin, P., Choi, H., Calvert, C., Pierce, N.: Programming biomolecular self-assembly pathways. Nature 451(7176), 318–322 (2008)
10. Yurke, B., Mills, A.: Using DNA to Power Nanostructures. Genetic Programming and Evolvable Machines 4(2), 111–122 (2003)
11. Zhang, D., Turberfield, A., Yurke, B., Winfree, E.: Engineering entropy-driven reactions and networks catalyzed by DNA. Science 318(5853), 1121 (2007)

Distributed Agreement in Tile Self-assembly

Aaron Sterling*

Department of Computer Science, Iowa State University
sterling@cs.iastate.edu

Abstract. Laboratory investigations have shown that a formal theory of fault-tolerance will be essential to harness nanoscale self-assembly as a medium of computation. Several researchers have voiced an intuition that self-assembly phenomena are related to the field of distributed computing. This paper formalizes some of that intuition. We construct tile assembly systems that are able to simulate the solution of the wait-free consensus problem in some distributed systems. This potentially allows binding errors in tile assembly to be analyzed (and managed) with positive results in distributed computing, as a "blockage" in our tile assembly model is analogous to a crash failure in a distributed computing model. We also define a strengthening of the "traditional" consensus problem, to make explicit an expectation about consensus algorithms that is often implicit in distributed computing literature. We show that solution of this strengthened consensus problem can be simulated by a two-dimensional tile assembly model only for two processes, whereas a three-dimensional tile assembly model can simulate its solution in a distributed system with any number of processes.

1   Introduction

One emerging field of computer science research is the algorithmic harnessing of molecular self-assembly to produce structures (and perform computations) at the nano scale. In his Ph.D. thesis in 1998, Winfree [16] used tiles on the integer plane to define a self-assembly model which has become an influential tool. As noted by several researchers (for example [1]), problems in algorithmic tile self-assembly share characteristics with better-studied problems in distributed computing: asynchronous computation, the importance of fault tolerance, and the limitations of local knowledge. In this paper, we formalize a connection between the two fields, by constructing models of tile assembly that simulate solutions to the wait-free consensus problem in some distributed systems.

The tile self-assembly literature has considered two main classes of models: models in which tiles bind to one another in an error-free manner, and models in which there is a positive probability that mismatched tiles will bind to one another. In an error-permitting model, if mismatched tiles bind, they can produce a *blockage*—an unplanned tile configuration that stops a particular section of an assembly from being able to accrete tiles.

* This research was supported in part by National Science Foundation Grants 0652569 and 0728806.

R. Deaton and A. Suyama (Eds.): DNA 15, LNCS 5877, pp. 154–163, 2009.

As blockages do occur in wetlab self-assembly experiments, it is natural to ask how we could make our self-assembly computations as resilient against blockages as possible. Researchers have investigated mechanisms to limit the chance of blockages through error-correction (for example [3] [5]), or, relatedly, for a tile assembly to "heal" itself in the event of damage [12]. Like other error-correcting codes, these mechanisms can consume significant overhead, and only reduce without eliminating the possibility of a blockage. Our interest in this paper, therefore, is to build a framework for robust self-assembly even in the presence of one or more unhealable blockages. Of course, if we consider a situation in which multiple subassemblies grow independently, then the blockage of one subassembly will have no effect on the others. The problem arises when otherwise independent subassemblies send information to, or receive information from, one another, and need to coordinate based on that information—hence our motivation to import the consensus problem into the world of tile assembly.

The most common types of processor faults modeled in distributed computing are *crash failure* (where a processor stops functioning) and *Byzantine failure* (where a processor can behave maliciously and take "worst-possible" steps). Several other types of failure have been defined; in general, their severity lies between crash failure and Byzantine failure. We will focus on shared objects that can be simulated in the presence of a tile self-assembly analogue of crash failures, to construct a theoretical foundation for synchronized fault-tolerance in self-assembly. In the long run, we believe that the combination of error-correction and distributed computing techniques (to manage a variety of failures) will produce self-assembling systems with high fault-tolerance.

The consensus problem was originally defined by Lamport for a system of distributed processors, as an abstraction of the transaction commit problem in database theory. It has since been shown to have wide application to the study of distributed systems; see, for example, Attiya and Welch [2] for a textbook introduction. In brief, given a system of n distributed processors, a *solution to the consensus problem* is an algorithm that ensures all nonfaulty processors agree on the same value. (There is also a "validity" condition to ensure the algorithm is not trivial.) The consensus problem for a system of n processors is called n-consensus, and a consensus algorithm is termed *wait-free* if up to $n-1$ processors can crash in an n processor system, and even so all correctly working processors will decide on the same value.

This paper is part of a larger program to connect the fields of self-assembly and distributed computing. By reducing models of self-assembly to models of distributed processors, one can apply known distributed computing impossibility results to obtain limits to the power of self-assembly [13] [14]. On the other hand, by simulating models of distributed computing in self-assembly, one can use strong techniques like ultrametrics to generalize known self-assembly results [15]. The objective of the current paper is to explore which distributed objects can (and cannot) be simulated by self-assembling systems, in order to clarify how positive results of distributed computing can apply to self-assembly.

We have organized the rest of the paper as follows. Section 2 provides further background about tile self-assembly and distributed computing. In Section 3, we construct a tile assembly simulation of wait-free 2-consensus. In Section 4, we define a strengthening of the consensus problem, and show that two-dimensional tile assembly systems cannot simulate solutions to it for systems of three or more processes, but three-dimensional tile assembly systems can. Section 5 concludes the paper and provides suggestions for future research.

2   Background

2.1   Tile Self-assembly Background

Winfree's objective in defining the Tile Assembly Model was to provide a useful mathematical abstraction of DNA tiles combining in solution in a random, non-deterministic, asynchronous manner [16]. Rothemund [10], and Rothemund and Winfree [11], extended the original definition of the model. For a comprehensive introduction to tile assembly, we refer the reader to [10]. Intuitively, we desire a formalism that models the placement of square tiles on the integer plane, one at a time, such that each new tile placed binds to the tiles already there, according to specific rules. Tiles have four sides (often referred to as north, south, east and west) and exactly one orientation, *i.e.*, they cannot be rotated.

A tile assembly system T is a 5-tuple $(T, \sigma, \Sigma, \tau, R)$, where T is a finite set of tile types; σ is the *seed tile* or *seed assembly*, the "starting configuration" for assemblies of T; $\tau : T \times \{N, S, E, W\} \rightarrow \Sigma \times \{0, 1, 2\}$ is an assignment of symbols ("glue names") and a "glue strength" (0, 1, or 2) to the north, south, east and west sides of each tile; and a symmetric relation $R \subseteq \Sigma \times \Sigma$ that specifies which glues can bind with nonzero strength. In this model, there are no negative glue strengths, *i.e.*, two tiles cannot repel each other.

In this paper, we allow for the possibility of errors in binding between tiles. While, in general, binding errors can cause unplanned configurations to be built, we will make a simplifying assumption that the only binding errors that might occur are *tile blockages*, tile mismatches that prevent any further tiles from binding to the subassembly at which the blockage occurred. In particular, no erroneously bound tile can be enclosed by tiles that attach later in the process of self-assembly.

A *configuration of* T is a set of tiles, all of which are tile types from T, that have been placed in the plane, and the configuration is *stable* if the binding strength (from τ and R in T) at every possible cut is at least 2. An *assembly sequence* is a sequence of single-tile additions to the frontier of the assembly constructed at the previous stage. Assembly sequences can be finite or infinite in length. The *result* of assembly sequence $\overrightarrow{\alpha}$ is the union of the tile configurations obtained at every finite stage of $\overrightarrow{\alpha}$. The *assemblies produced by* T is the set of all stable assemblies that can be built by starting from the seed assembly of T and legally adding tiles. If α and β are configurations of T, we write $\alpha \longrightarrow \beta$ if there is an assembly sequence that starts at α and produces β. An assembly of T is *terminal* if no tiles can be stably added to it.

2.2   Distributed Computing Background

Distributed computing began as the study of networks of processors, in which each processor had limited local knowledge. However, much of the distributed computing literature now speaks in terms of systems of *processes*, not processors, to emphasize that the algorithms or bounds obtained from the theorem apply to any appropriate collection of discrete processes, whether they are part of the same multicore chip, or spread across a sensor network.

One of the most-studied distributed computing models is the *asynchronous shared memory model*, in which processes are modeled as finite state machines that can read and write to one or more shared memory locations called *registers*. The model is asynchronous because there is no restriction on when a process will execute its next computation step, except that any nonfaulty process can delay only a finite length of time before taking its next step. We refer the reader to [2] for exposition and technical details of such a model. We now provide a formal definition of the consensus problem for a distributed system.

Definition 1. *Let $\mathcal{M}$ be an asynchronous shared memory model such that each process p_i in $\mathcal{M}$ has two special state components: x_i, the input; and y_i, the decision value. Let V, the set of possible decision values, be a finite set of positive integers. We require that $x_i \in V$ for all i. For each i, y_i starts out containing a null entry, y_i is write-once, and the value written cannot be erased. A solution to the consensus problem is an algorithm that guarantees the following.*

Termination. *Every nonfaulty process p_i eventually writes a value to y_i.*
Agreement. *For any i, j, if p_i and p_j are nonfaulty and write to y_i and y_j, then $y_i = y_j$. All nonfaulty processes decide on the same decision value.*
Validity. *If all processes hold the same input value at the start of execution of the algorithm, then any value decided on must be that common input value.*

We will consider *fault-tolerant consensus*, in which one or more processes can fail in some way. The simplest type of failure—and the only type we will consider in this paper—is *crash failure*, meaning that at some point a process ceases operation and never takes another step.

It is a classic result of distributed computing that there is no deterministic algorithm that solves the consensus problem in an asynchronous shared memory model, in the presence of even a single crash failure, when the only registers available to the system are read/write registers [4]. One way to overcome that impossibility result is by making the registers more powerful. These more powerful registers are often called *objects* or *shared objects*, to emphasize they might be a special process segment unto themselves, not just a single memory location. For a comprehensive introduction to the theory of shared objects with different consensus strengths, see [7]. We will use only the following key definitions in this paper.

Definition 2. *Object O solves wait-free n-process consensus if there exists an asynchronous consensus algorithm for n processes, up to $n-1$ of which might fail (by crashing), using only shared objects of type O and read/write registers. In* *distributed system* $\mathcal{M}$, O *is a* consensus object *if each process in* $\mathcal{M}$ *can invoke* O *with a command* **invoke**(O,v), *where* $v \in V$ *is a possible decision value, and* O *will ack with command* **return**(O, v_{out}), *where* $v_{out} \in V$ *is a possible decision value, and* O *returns the same value* v_{out} *to all processes that invoke it.* O *is an* n-consensus object *if* O *is a consensus object and* n *is maximal such that* O *solves wait-free* n-process consensus.

Finally, we define the notions of configurations and execution segments of a distributed system. Intuitively, we consider the events of a system to be the read and write invocations (and acks) performed upon (and returned by) registers by the processors of the system; and configurations and execution segments are built up from the instantaneous state of the system, and events that are then applied to it.

Definition 3. *Let* $\mathcal{M}$ *be a distributed system with* n *processes and one* n-consensus *object* O. *A* configuration *of* $\mathcal{M}$ *is an* $(n + 1)$-*tuple* $\langle q_1, \ldots, q_n, o \rangle$, *where* q_i *is a state of* p_i *and* o *is a state of* O. *The* events *in* $\mathcal{M}$ *are the computation steps by the processes, the transmission of a consensus decision value from* O, *and the (possible) crashes of up to* $n - 1$ *of the processes. A* legal execution segment *of* $\mathcal{M}$ *is a sequence of form* $C_0, \phi_0, C_1, \phi_1, C_2, \phi_2 \ldots$ *where each* C_i *is a configuration,* ϕ_i *is an event, and the application of* ϕ_i *to* C_{i-1} *results in* C_i.

3   Simulating 2-Consensus in Two Dimensions

We turn now to the simulation in tile assembly of shared objects that solve wait-free consensus problems. For clarity of exposition, we will discuss using figures how to simulate 2-consensus by a two-dimensional tile assembly system, and in the next section sketch how to extend the simulation to more processes, and to a third tiling dimension.

Winfree showed how to simulate the behavior of a Turing machine by means of a "wedge construction" in tile assembly [16]. Lathrop *et al.* [8] extended this wedge construction so that the notion of "simulate" would mean that each row of the wedge records the state of the Turing machine tape after one move of the head, and, if the Turing machine halts, a row of tiles is built along the side of the wedge, so a special "halting tile" binds to the base of the wedge, as a marker that the simulation has halted. Inspired by [8], we recently demonstrated a construction whereby tiles can simulate distributed processors in a message-passing system (Theorem 6 of [15]), and we will now modify that machinery, to construct a tile simulation of processors that communicate with shared objects. (For reasons of space, we will not formally define what it means for a wedge construction to "simulate" a distributed processor p, but the intuition is that we add an input buffer and an output buffer to Winfree's original construction by having every other row of the wedge check to see whether a message has been received, and to incorporate the message if one has arrived; and we simulate the sending and receipt of a constant-size number of messages by using a unique tile that encodes the message to build a ray along the edge of the wedge, toward the intended destination of the message, and a ray back along the edge of the wedge to simulate the *ack*—the character that acknowledges receipt of the message.)

Intuitively, the main objective of this section is to construct a tile assembly system that behaves as shown in Figure 1: there are three "modules" or sub-assemblies: one (π_1) to simulate p_1, another (π_2) to simulate p_2, and a third to simulate a 2-consensus object. In addition, there are "rays" of tiles built from π_1 and π_2, to simulate the values that p_1 and p_2 are writing to the 2-consensus object; and "rays" of tiles built in response, to simulate the acks received by p_1 and p_2 after the writes conclude. The main formal objective in this section is to sketch the proof of Theorem 1, stated below.

Definition 4. *Tile assembly system $\mathcal{T}$ simulates distributed system $\mathcal{M}$ if:*

1. *There is a 1-1 mapping h from configurations of $\mathcal{M}$ to stable tile configurations of $\mathcal{T}$.*
2. *If $C_0, \phi_0, C_1, \phi_1 \cdots C_i, \phi_i$ is a legal execution segment of $\mathcal{M}$, then $h(C_0) \longrightarrow h(C_1) \longrightarrow \cdots \longrightarrow h(C_i)$ is a legal tile assembly sequence in $\mathcal{T}$.*
3. *If there is no legal execution segment from C_0 to C_1 in $\mathcal{M}$, then there is no legal tile assembly sequence in $\mathcal{T}$ such that $h(C_0) \longrightarrow h(C_1)$.*
4. *Let C be a configuration of $\mathcal{M}$ and $\mathcal{C}$ be a set of configurations of $\mathcal{M}$. If $\mathcal{M}$ is such that, upon achieving configuration C, it must eventually achieve some configuration $C' \in \mathcal{C}$ unless a process crashes, then $\mathcal{T}$ is such that, if it ever reaches $h(C)$ then it must achieve $h(C')$ for some $C' \in \mathcal{C}$ unless there is a tile blockage. (Note that $\mathcal{C}$ may be an infinite set.)*
5. *If C_0, ϕ_0, C_1, ϕ_1 is a legal execution segment in $\mathcal{M}$, and the event ϕ_0 is the crash failure of a previously correct process, then $h(C_1)$ contains one more tile blockage binding error than $h(C_0)$ contains.*
6. *If in configuration C of $\mathcal{M}$ all processes have halted, then in tile configuration $h(C)$ of $\mathcal{T}$ there are no further locations to which tiles can bind stably, i.e., $h(C)$ is terminal.*

Theorem 1. *Let $\mathcal{M}$ be an asynchronous message passing model of distributed computing with two processes and one 2-consensus object O, such that p_1 and p_2 send no messages to each other, and such that each process invokes O at most once. Then there is a tile assembly system $\mathcal{T}$ that simulates $\mathcal{M}$.*

To prove Theorem 1 we first exhibit a tile configuration that can simulate a 2-consensus object.

Lemma 1. *Fix V any finite set. There is a tile configuration ρ that contains a binding location l with the following properties: (1) the only tiles that bind at l have nonzero glue strengths at either the south, north and west sides, or the south, north and east sides; (2) any tile that binds at l will have a glue name taken from V on either its east or west side; (3) if glue name $v \in V$ is on the tile that binds at l, then the name of the tile's north glue will be "Ackv," and ρ will build a ray transmitting the glue name Ackv to its west and east.*

The proof of Lemma 1 formalizes the construction shown in Figure 2, and the proof of Theorem 1 formalizes the construction shown in Figure 1.

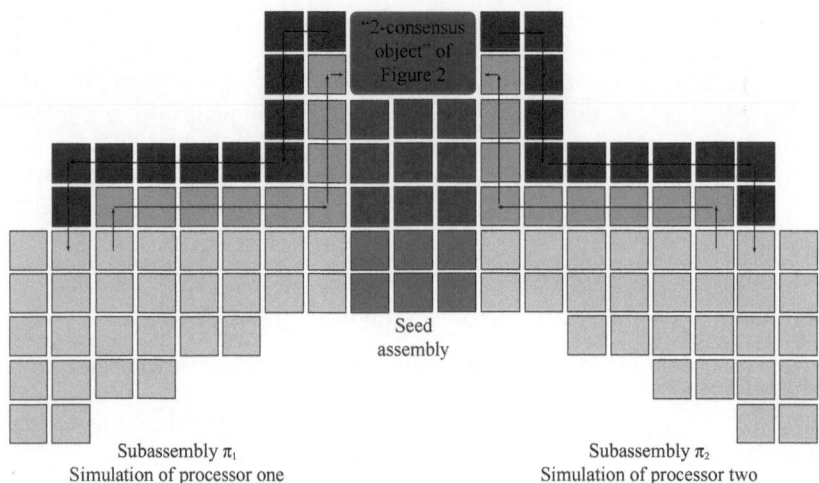

Fig. 1. Two (yellow) subconfigurations π_1 and π_2 growing from a common (green) seed assembly. (The arrows indicate the order in which tiles bind to the assembly.) They communicate with the configuration shown in Figure 2, by means of (blue) "message" tiles they send, and (purple) "ack" tiles they receive back. Intuitively, π_1 and π_2 simulate two processes in a distributed system, communicating with a 2-consensus object.

4   Simulating Distributed Systems with Three or More Processes

One critical difference between "classical" distributed systems and tile assembly systems is that sending messages—and writing to shared memory locations—in a distributed system does not affect the future computation resources available to processes; whereas in a tile assembly system, the tiles placed on the plane to simulate such operations may "box in" other subassemblies, so they cannot grow beyond some point, due to tile blockages. Put another way, systems of distributed processes have multidimensional resources: each process computes using its own set of resources, and message-passing takes place via a different set of resources. By contrast, every tile operation self-assembles using the same shared resource: the surface. It is, therefore, not surprising that to simulate this resource-independence of distributed systems, tile assembly systems require multiple surfaces, that is to say, three spatial dimensions. Formalizing that is the main objective of this section.

Definition 5. *Let $\mathcal{M}$ be a system of distributed processes such that all correctly-operating processes will run forever. A solution to the* Consensus Subroutine Problem *for $\mathcal{M}$ is an algorithm A that solves the wait-free consensus problem for $\mathcal{M}$ in such a way that no process increases its likelihood of crash failure by calling A as a subroutine.*

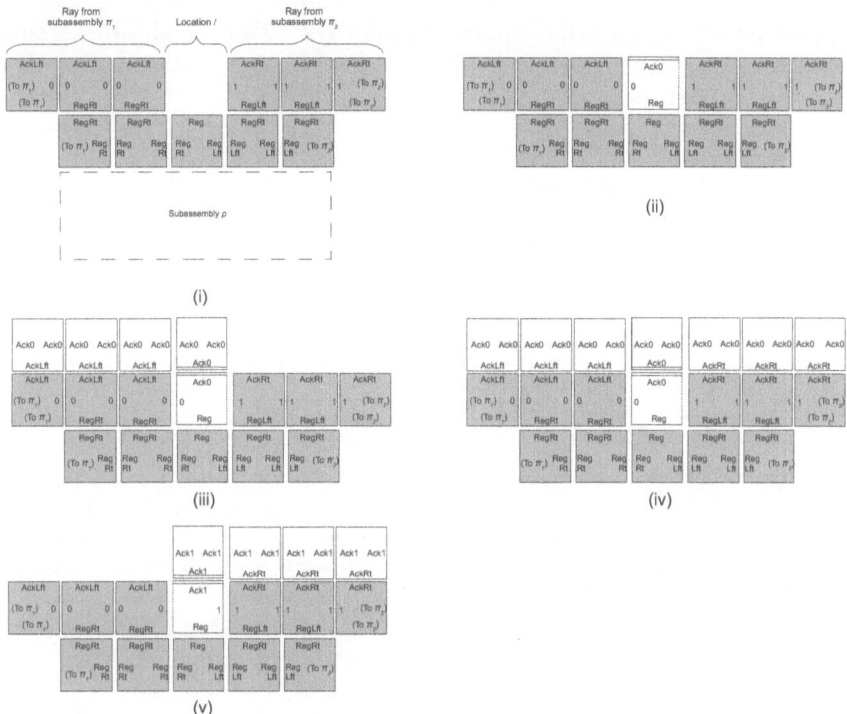

Fig. 2. Diagram of the simulator of a 2-consensus object used to prove Theorem 1. At stage (i), rays from subassemblies π_1 and π_2 approach location l. At stage (ii), a tile binds at l, in this case deciding in favor of the input value of subassembly π_1. At stage (iii), the simulator sends an ack to π_1. At stage (iv), the simulator sends an ack to π_2. "Stage" (v) demonstrates the alternative: a decision tile has bound in favor of the initial value of π_2, and the simulator has acked to π_2.

If $\mathcal{M}$ has only two processes, then a tile assembly system of the form in Figure 1 will solve the Consensus Subroutine Problem for $\mathcal{M}$, because there is enough space on the surface for π_1, π_2 and the messages to and from ρ to assemble without interfering. For $\mathcal{M}$ with three or more processes, the situation is different. If we attempt to simulate processes that run forever in two dimensions, the tile configurations that simulate the processes must grow without bound. This will cause destructive collision with tiles attempting to communicate information with an n-consensus object.

Theorem 2. *Let $\mathcal{M}$ be a system of n distributed processes ($n \geq 3$) and one n-consensus object, such that all correctly-working processes will run forever. Then no two-dimensional tile assembly system can simulate a solution of the Consensus Subroutine Problem for $\mathcal{M}$.*

We omit the proof. The main idea is that if three independent subassemblies each wish to build rays toward a single point, there exist worst-case assembly sequences in which one subassembly can only obtain the information by building a ray that blocks the growth of another subassembly. Thus, to ensure the simulation is wait-free for all possible assembly sequences, some subassembly cannot run forever, in contradiction to the hypothesis.

Theorem 3. *Let M be a system of n distributed processes $(n \geq 3)$ and one n-consensus object O, such that the processes do not send one another messages, each process invokes O at most once, and all correctly-working processes will run forever. There is a three-dimensional tile assembly system that simulates solution of the Consensus Subroutine Problem for M.*

We omit the proof. It hinges on the fact that three dimensions provides us enough space to build rays to and from each independent subassembly so that we can generalize the construction of Figure 2.

5   Conclusion

We have shown how two-dimensional tile assembly systems can simulate solution to the consensus problem for some two-process distributed systems, and how three-dimensional tile assembly systems can simulate a strengthening of the consensus problem for some n-process distributed systems, for any n. We only simulated systems in which the processors do not communicate with one another. One way to extend our current results would be to consider what types of communication among processes could be simulated by tile assembly.

In this paper, we assumed that the only way a tile simulation could fail was via a blockage that immediately caused the entire subassembly to stop growing. A more extreme failure assumption (analogous to "Byzantine" failures in the language of distributed computing) would be that a subassembly might grow in a malformed, haywire fashion. It would be interesting to see what application research into Byzantine failures might have to issues of fault-tolerance in tile self-assembly.

Acknowledgements

I am grateful to Jim Lathrop, Jack Lutz and Scott Summers for helpful discussions on earlier versions of this paper. I am especially grateful to Soma Chaudhuri for many helpful discussions.

References

1. Arora, S., Blum, A., Schulman, L., Sinclair, A., Vazirani, V.: The computational worldview and the sciences: a report on two workshops. NSF Report (October 2007)
2. Attiya, H., Welch, J.: Distributed Computing: Fundamentals, Simulations, and Advanced Topics, 2nd edn. Wiley Series on Parallel and Distributed Computing (2004)

3. Chen, H.-L., Goel, A.: Error free self-assembly using error prone tiles. In: Ferretti, C., Mauri, G., Zandron, C. (eds.) DNA 2004. LNCS, vol. 3384, pp. 62–75. Springer, Heidelberg (2005)
4. Fischer, M., Lynch, N., Paterson, M.: Impossibility of Distributed Consensus with One Faulty Process. Journal of the ACM 32(2), 374–382 (1985)
5. Fujibayashi, K., Zhang, D.Y., Winfree, E., Murata, S.: Error suppression mechanisms for DNA tile self-assembly and their simulation. Natural Computing (published online July 9, 2008)
6. Herlihy, M.: Wait-free synchronization. ACM Transactions on Programming Languages and Systems 13(1), 124–149 (1991)
7. Herlihy, M., Shavit, N.: The Art of Multiprocessor Programming. Morgan Kaufmann, San Francisco (2008)
8. Lathrop, J., Lutz, J., Patitz, M., Summers, S.: Computability and complexity in self-assembly. In: Beckmann, A., Dimitracopoulos, C., Löwe, B. (eds.) CiE 2008. LNCS, vol. 5028, pp. 349–358. Springer, Heidelberg (2008)
9. Lathrop, J., Lutz, J., Summers, S.: Strict self-assembly of discrete Sierpinski triangles. In: Cooper, S.B., Löwe, B., Sorbi, A. (eds.) CiE 2007. LNCS, vol. 4497, pp. 455–464. Springer, Heidelberg (2007)
10. Rothemund, P.W.K.: Theory and Experiments in Algorithmic Self-Assembly. Ph.D. thesis, University of Southern California, Los Angeles (2001)
11. Rothemund, P., Winfree, E.: The program-size complexity of self-assembled squares. In: Proceedings of the 32nd Annual ACM Symposium on Theory of Computing, pp. 459–468 (2000)
12. Soloveichik, D., Cook, M., Winfree, E.: Combining Self-Healing and Proofreading in Self-Assembly. Natural Computing 7(2), 203–218 (2008)
13. Sterling, A.: A limit to the power of multiple nucleation in self-assembly. In: Taubenfeld, G. (ed.) DISC 2008. LNCS, vol. 5218, pp. 451–465. Springer, Heidelberg (2008)
14. Sterling, A.: A limit to the power of multiple nucleation in self-assembly (full version) (submitted), http://arxiv.org/abs/0902.2422v1
15. Sterling, A.: Brief announcement: self-assembly as graph grammar as distributed system. To appear in: Proceedings of the 28th Annual ACM SIGACT-SIGOPS Symposium on Principles of Distributed Computing (2009)
16. Winfree, E.: Algorithmic Self-Assembly of DNA. Ph.D. thesis, California Institute of Technology, Pasadena (1998)

Author Index